T0227433

Advances in Soil Science

CROPS RESIDUE MANAGEMENT

Advances in Soil Science

CROPS RESIDUE MANAGEMENT

Edited by

J.L. Hatfield
B.A. Stewart

CRC Press
Taylor & Francis Group
Boca Raton London New York

CRC Press is an imprint of the
Taylor & Francis Group, an **informa** business

First published 1994 by CRC Press
Taylor & Francis Group
6000 Broken Sound Parkway NW, Suite 300
Boca Raton, FL 33487-2742

Reissued 2018 by CRC Press

Library of Congress Cataloging-in-Publication Data

Catalog record is available from the Library of Congress

A Library of Congress record exists under LC control number: 93038816

Publisher's Note
The publisher has gone to great lengths to ensure the quality of this reprint but points out that some imperfections in the original copies may be apparent.

Disclaimer
The publisher has made every effort to trace copyright holders and welcomes correspondence from those they have been unable to contact.

ISBN 13: 978-1-315-89214-6 (hbk)
ISBN 13: 978-1-351-07124-6 (ebk)

Visit the Taylor & Francis Web site at http://www.taylorandfrancis.com and the
CRC Press Web site at http://www.crcpress.com

Preface

Conservation tillage and residue management have gained popularity in the past 5 years due to the provisions in the 1985 Food Security Act and the impetus to reduce production costs by the farmers. Conservation tillage has been defined as any practice which leases more than 30% cover on the soil surface after planting, while residue management is more of a concept or a strategy directed toward the efficient management of the residue to reduce soil erosion and improve soil quality. Often the terms are used interchangeably; however, there are differences which must be understood.

The Conservation Technology Information Center (CTIC) has shown that there is an increasing trend in the use of conservation tillage methods. In their 1992 report there was a name change from the *National Survey of Conservation Tillage Practices* to *National Crop Residue Management Survey*. This survey reported three levels of residue: 0-15%, 15-30%, and greater than 30%. In 1992 the use of conservation tillage increased to 31.2% on all U.S. acres up from 28.1% in 1991. Residue management programs which maintain residue levels above 15% have continued to increase and these increases have come from fields which were traditionally in the low residue category. Changes are occurring in fields across the United States and Canada and the changes are directed toward managing the crop residue to decrease water and wind erosion and increase soil quality. Farmers are adopting technologies and modifying equipment to place fertilizer near the seed, move residue away from the seed zone without burying the residue, modify planting equipment to work with residue, and try different weed control strategies. Farmers' needs for for information have shown where some of the deficiencies are in our understanding of how crop residue can be effectively managed.

In 1992 a workshop was held in Kansas City, Missouri to evaluate the current state of knowledge regarding residue management strategies. This workshop was attended by over 175 scientists, extension specialists, and soil conservationists who over a 2 1/2 day period discussed what we know about residue management and proposed areas where additional information is needed. Only a portion of that workshop is contained in this volume. The chapters prepared by a range of experts for this volume represent the general information and techniques used for residue management. Another product forthcoming from this workshop will be a series of technology transfer documents that summarize the available information and concepts for each region of the United States.

This volume was organized to show the residue management strategies in different regions of the United States and to summarize where the current limitations are in applying residue management concepts. The areas of the United States are the Northeast, Pacific West, Great Plains, Midwest, and Southeast. Each of these areas has unique characteristics and responses to residue management. The remaining chapters detail the principles involved in residue management, effects on soil erosion, effects on weeds, effects of cover crops on crop residue, and the soil, climate, residue interactions. There is a large body of scientific knowledge; however, we have not begun to develop this information into general principles which can be adapted and applied throughout the United States or the world.

The challenge for the reader is to begin to extend this information and knowledge into management strategies which will improve the effectiveness of crop residue management. It is our hope that each reader will be inspired to work with other scientists to improve our understanding of what crop residue does to the soil and to develop information which can be applied by the farmer and the conservationist.

J.L. Hatfield
B.A. Stewart

Contributors

E.E. Alberts, U.S. Department of Agriculture, Agricultural Research Service, Cropping Systems and Water Quality Research Unit, Columbia, MO 65211, U.S.A.

R.L. Blevins, Department of Agronomy, University of Kentucky, Lexington, KY 40446, U.S.A.

R.R. Bruce, U.S. Department of Agriculture, Agricultural Research Service, Southern Piedmont Conservation Research Unit, Watkinsville, GA 30677, U.S.A.

D.D. Buhler, U.S. Department of Agriculture, Agricultural Research Service, Plant Science Research Unit, St. Paul, MN 55108, U.S.A.

W.M. Edwards, U.S. Department of Agriculture, Agricultural Research Service, North Appalachian Experiment Watershed, Coshocton, OH 43812, U.S.A.

F. Forcella, U.S. Department of Agriculture, Agricultural Research Service, North Central Soil Conservation Research Laboratory, Morris, MN 56267, U.S.A.

W.W. Frye, Department of Agronomy, University of Kentucky, Lexington, KY 40466, U.S.A.

D.R. Griffith, Department of Agronomy, Purdue University, West Lafayette, IN 47907, U.S.A.

C.W. Honeycutt, U.S. Department of Agriculture, Agricultural Research Service, New England Plant, Soil, and Water Laboratory, Orono, ME 04469, U.S.A.

G.W. Langdale, U.S. Department of Agriculture, Agricultural Research Service, Southern Piedmont Conservation Research Unit, Watkinsville, GA 30677, U.S.A.

D.K. McCool, U.S. Department of Agriculture, Agricultural Research Service, Land Management and Water Conservation Research Unit, Pullman, WA 99164-6421, U.S.A.

M.E. McGiffen, Department of Botany and Plant Sciences, University of California, Riverside, CA 92521, U.S.A.

K.C. McGregor, U.S. Department of Agriculture, Agricultural Research Service, National Sedimentation Laboratory, Oxford, MI 38655, U.S.A.

R.I. Papendick, U.S. Department of Agriculture, Agricultural Research Service, Land Management and Water Conservation Research Unit, Pullman, WA 99164-6421, U.S.A.

J.K. Radke, U.S. Department of Agriculture, Agricultural Research Service, National Soil Tilth Laboratory, Ames, IA 50011, U.S.A.

D.W. Reeves, U.S. Department of Agriculture, Agricultural Research Service, National Soil Dynamics Laboratory, Auburn, AL 36831-3439, U.S.A.

D.C. Reicosky, U.S. Department of Agriculture, Agricultural Research Service, North Central Soil Conservation Research Laboratory, Morris, MN 56367, U.S.A.

D.D. Tyler, Department of Plant and Soil Science, University of Tennessee, West Tennessee Experiment Station, Jackson, TN 38301-3200, U.S.A.

P.W. Unger, U.S. Department of Agriculture, Agricultural Research Service, Conservation and Production Research Laboratory, Bushland, TX 79012, U.S.A.

M.G. Wagger, Department of Soil Science, North Carolina State University, Raleigh, NC 27695, U.S.A.

N.C. Wollenhaupt, Department of Soil Science, University of Wisconsin, Madison, WI 53706, U.S.A.

Contents

Residue Management Strategies – Pacific Northwest

R.I. Papendick and D.K. McCool

I. Introduction

Crop residue is a major tool for controlling erosion on Northwest wheatlands and is included in virtually all farm plans to achieve conservation compliance. Crop residues also conserve water by reducing runoff and evaporation which is paramount for economic crop production in the whole region. The principal dry-farming area consists of 3.3 million ha of croplands in a contiguous belt extending across northern Idaho, eastern Washington, and north central Oregon (Thomas and Grano, 1971). This region is most known for its production of soft, white winter wheat (*Triticum aestivum* L.) and the food legumes pea (*Pisum sativum*) and lentil (*Lens culinaris*).

II. Description of the Region

A. Climate

The climate varies from semiarid at the western edge to subhumid approaching the mountainous areas to the east. Land elevation ranges from 350 to 1400 m (Austin, 1965). The entire region has wet winters and dry summers. Annual precipitation ranges from 200 to 600 mm of which 60 to 70% occurs during November through April (Ramig et al., 1983). Snowfall is 20 to 25% of the total precipitation at the higher elevations but decreases in the lower areas where it is drier. Soil freezing may occur to depths ranging from a few to 10 cm several times each winter and to 40 cm or more during some winters. These events are often interrupted by partial or complete thaws caused by warm frontal systems with rainy weather from the Pacific Ocean. Though the rainfall intensities are usually less than 4 mm/hr (Horner et al., 1957), considerable runoff can occur while the soils are frozen, especially if they are initially snowcovered.

The region is also subject to high winds in the spring and fall which create a wind erosion hazard in the drier areas where soils are relatively sandy and cover is sparse. Because of the rolling topography, the wind speed profiles are likely to be different from flat terrain and produce turbulence effects on different parts of the landscape. Heretofore, wind erosion has not been studied in the region.

B. Topography and Soils

The topography in the higher precipitation areas is steeply rolling with dune-like hills and steep north and east slopes. Most slopes range from 8 to 30% in steepness but slopes in excess of 45% are cultivated (Ramig et al., 1983). Most fields classify as highly erodible land. In the drier areas the topography is gently rolling and less land is classified as highly erodible for water erosion and more for wind erosion. Throughout the region, narrow valleys together with the slope pattern form a well-defined drainage system.

Soils are derived from loess mixed with varying amounts of volcanic ash. They are classified as Mollisols, sub-order Xerolls and great groups Argixerolls and Haploxerolls (U.S. Geological Survey, 1970). When not influenced by a restrictive frost layer, the soils are generally permeable, well-drained and can store the annual precipitation. Some soils in north central Oregon are shallow and have limited storage capacity. Soils tend to be sandy in the drier areas and grade to silt loams and silty clay loams in areas with increasing precipitation.

C. Cropping Practices

Cropping systems depend primarily on the annual precipitation. The region's croplands can be grouped into three precipitation zones. These are high (more than 380 mm), intermediate (330 to 380 mm) and low (200 to 330 mm) (Ramig et al., 1983). In the high precipitation zone annual cropping is practiced and winter wheat is grown in rotation with pea, lentil, spring barley (*Hordeum vulgare*) and spring wheat. Based on discussions with USDA Soil Conservation Service field personnel and a survey of farms in the area, crop percentages are winter wheat 50%, pea and lentil 20%, spring barley and wheat 20%, and other (set aside grass, fallow) 10 %. The usual cropping system in the intermediate zone is winter wheat-spring barley or wheat-fallow. The crop percentages are winter wheat 45%; spring barley and wheat 20%; and fallow and other 35%. In the low precipitation zone (approximately one-half of the total cropland area) it is 50% each for winter wheat and fallow.

Winter wheat in the high precipitation zone produces from 7 to 13 t ha^{-1} crop residue and 3 to 6 t ha^{-1} in the low precipitation zone. Stubble produced by spring barley and wheat ranges from 0.5 to 0.7 of that produced by winter wheat. Pea and lentil residue is generally no more than 3 t ha^{-1} at harvest.

In the higher precipitation zones where winter wheat is planted in October there is very little crop growth in the fall. Winter wheat is essentially dormant from December until March and most growth, as with spring crops, occurs from April to mid-July on residual soil moisture. In the fallow areas where crops are planted earlier there is considerably more growth in the fall (in some cases complete ground cover before winter) and growth resumes earlier in the spring.

III. Erosion and Water Conservation Problems

The climatic pattern, steep topography and winter wheat cropping with conventional tillage creates a winter runoff and water erosion problem in much of the region. It is not uncommon to have water erosion rates of 100 to 250 t ha^{-1} on some slopes and field averages of 20 to 40 t ha^{-1} in a single season (Papendick et al., 1983). Erosion rates are highest when the soils have been frozen and are partially thawed (Horner et al., 1944). Frozen soil runoff is closely correlated with snow cover and amount of rainfall because infiltration is very slow (Horner et al., 1944; Zuzel and Pikul, 1987). For example, runoff from untilled stubble may be greater than from bare soil because the stubble can hold more drifting snow. However, the converse sometimes occurs because 15 to 20 cm of snow insulates the soil enough to prevent freezing.

Wind erosion rates can exceed those by water. The first author personally observed fields in south central Washington that lost between 4 to 10 cm of topsoil during fall to spring in 1990-91 from wind erosion.

Evaporation is the major source of water loss. It is estimated that 75% of the annual precipitation can be lost by evaporation from a bare, uncropped soil (Bristow et al., 1986). Although potential evaporation is greatest during the summer, the largest actual loss occurs by first stage drying during the winter rainy season and in the early spring as radiation increases and the soil surface is still moist (Papendick and Campbell, 1974).

IV. Effectiveness of Residues for Erosion Control

The benefits of surface residue for erosion control are well established and effectiveness relationships have been developed for Pacific Northwest conditions. Effects of incorporated residue are not as well developed but a relationship has been established for use in the Revised Universal Soil Loss Equation (Yoder et al., 1994).

A. Surface Residues

In the Pacific Northwest the main effect of surface residue on erosion is in reducing the transport capacity of runoff water (McCool et al., 1977). The effect of residue cover on the soil erosion is given by the equation:

$$SC = \exp(-B*RC) \tag{1}$$

where SC is the residue cover subfactor, B is a coefficient and RC is percent of the soil surface covered by the residue (Yoder et al., 1994). The residue cover subfactor is the fraction of soil loss for a given percent cover as compared with no cover. Experimentally determined B values for the Pacific Northwest are generally greater than 0.050 compared with an average of 0.035 for the Midwest. The relationships graphed in Figure 1 show residue cover is highly effective for erosion control and more so in the Pacific Northwest than in the Midwest. The reason for this is that rill erosion, which is the dominant process in the Pacific Northwest, has higher B values than mixed interrill/rill erosion which is the dominant process in the Midwest.

The relationship between residue cover in percent and mass is given by the equation:

$$RC = [1-\exp(-A*RW)]*100 \tag{2}$$

where RC is the percent residue cover, A is the ratio of the area covered by the residue to the mass of the residue, and RW is the residue mass per unit area (Gregory, 1982). In a study of the characteristics of a one-year sampling of wheat, barley and pea stubble from eastern Washington, McCool et al. (1990) found values of A ranging from 0.00045 to 0.00081 ha/kg. At 50% cover, the

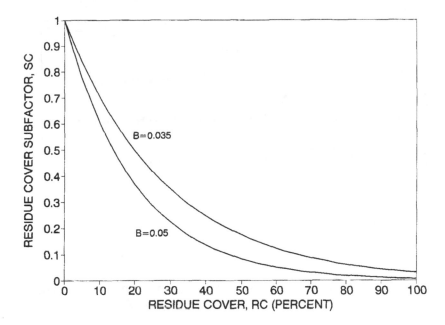

Figure 1. Effect of residue cover and B values in equation [1] on soil loss. (From Yoder et al., 1994.)

mass per unit area ranged from 850 to 1560 kg/ha, a ratio of nearly 2 to 1. Given a lack of specific varietal and yearly data, it is suggested that this range in A values could be adequately represented by three values: 0.00045 for large stemmed winter wheat, 0.00057 for medium stemmed winter wheat, peas and large stemmed winter barley, and 0.00080 for small stemmed spring barley varieties. Three curves representing these data are plotted in Figure 2.

B. Incorporated Residue

The effect on erosion of subsurface residue materials (including roots) from previous crops is given by:

$$PLU = DEN*exp(-C*RS) \qquad [3]$$

where PLU is the prior land use subfactor which ranges from 0 to 1, DEN is a surface soil consolidation factor, RS is the mass of roots and residue incorporated in the upper 10 cm of soil, and C is a coefficient expressing the effectiveness of the incorporated materials in controlling erosion (Yoder et al., 1994). DEN takes into account the effect of tillage-induced disruption of the surface layers and has values ranging from 1 for a freshly-tilled soil to less than

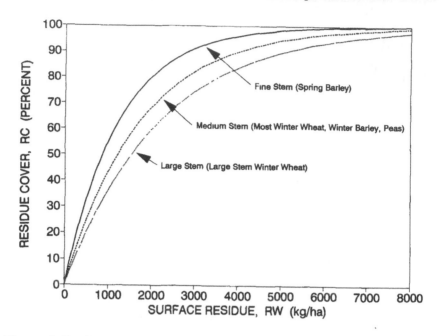

Figure 2. Residue cover vs. mass per unit area for typical Pacific Northwest nonirrigated crops. (From McCool et al., 1990.)

0.5 for undisturbed conditions. The value of C depends on the dominant erosion process. For rill erosion in the Pacific Northwest C has a value of 0.00040 ha/kg (Yoder et al., 1994). A graph of eq [3] using DEN of 1.0 illustrated in Figure 3 shows a significant benefit of buried residues for erosion control in the Pacific Northwest. A plotting of effectiveness of winter wheat residue versus surface mass per unit area using Pacific Northwest values of B and A in eqs [1] and [2] is included for comparison. The effect of incorporated residues is considerably less than that of surface residue.

V. Benefits of Residue for Water Conservation

Residues are most effective for water conservation during the first stage of evaporation when the soil surface is wet. This condition exists much of the winter and early spring in the Pacific Northwest which explains why surface mulches are successful in the region. Their value for water conservation is less during the summer when there is little precipitation. Papendick and Miller (1977) reported that with 0, 5, and 11 Mg ha^{-1} of residues on the surface from September through March water storage was 48, 66, and 81%, respectively, of the precipitation. In Oregon standing wheat stubble stored 87% of the precipitation from July 1 to April 1 whereas only 64% was stored where the

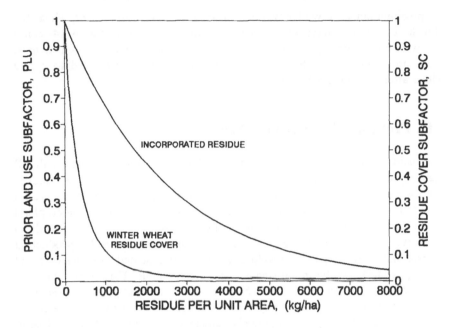

Figure 3. Effect of surface and incorporated residue on soil loss for typical Pacific Northwest nonirrigated conditions. (From Yoder et al., 1994.)

stubble was plowed after wheat harvest (Ramig and Ekin, 1976). The additional water from the stubble increased the yield of green peas 670 kg ha^{-1}. Gains in water storage with tillage treatments are generally greatest when surface residues are conserved (Papendick, 1987).

VI. The Situation on Farmers' Fields

To achieve conservation compliance many Pacific Northwest farmers must reduce water erosion on their highly erodible fields by 50 to 60 percent of levels using conventional management. Residue management must play an important role because the steep, irregular topography is not readily adaptable to other types of conservation measures. Residues are also critical for wind erosion control because the soils in the dry areas are poorly aggregated and do not form clods with tillage. Variable wind directions and landscape patterns may also limit the use of stripcropping and vegetative barriers as control practices.

The critical time of the crop rotation (also referred to as the critical year) for residue amount is when the fall crop is planted. In the high precipitation zone all fields coming out of pea, lentil and fallow are sown to winter wheat or occasionally to fall barley. A few fields out of spring wheat and winter wheat may also be fall planted to achieve the wheat base of the farm. In the intermediate precipitation zone winter wheat is sown after fallow and spring wheat but almost never after winter wheat because of the limited moisture. A 14-month fallow is the standard practice in the low precipitation zone after which most fields are planted to winter wheat. Occasionally some are planted to spring wheat after a 21-month fallow.

VII. Problems that Farmers Face

The main concerns with residue management are inadequate residues with fall planting following pea and lentil crops, and after fallow and excessive residues from winter wheat crops in the high and intermediate precipitation zones. A fall crop following pea, lentil and fallow represents over 80 percent of the cropland area sown to a fall crop each year. Inadequate residue cover can result in excessive erosion rates during the critical year. Besides low residue production by pea and lentil crops, the residues are more fragmented after harvest and decompose faster in the initial stages than cereal stubble. Farm plans that rely on residues may require from 30 to 50 percent cover at fall planting to achieve necessary erosion reductions. Any type of tillage followed by shank fertilizer application and seeding will commonly reduce surface cover of pea and lentil residue to 15 percent or less which is often an unacceptable level, especially with short rotations.

It is also difficult to maintain adequate residue cover with a year of tilled fallow. Residues are generally limited in the dry areas especially in years of below average crop yields. When any type of tillage system is used, especially fall tillage after harvest for weed control or to improve infiltration, the residue has largely disappeared by the time of planting.

To achieve compliance on a highly erodible field that is fallowed or planted into low residues every second or third year some farmers will maintain high residues on the field during the noncritical year. This reduces erosion to near zero during this period of the rotation and brings the rotational average for erosion on the highly erodible field to an acceptable level of predicted erosion control.

To accomplish this, a farmer in the high precipitation zone will leave wheat or barley stubble stand overwinter, or fall till with a chisel implement which leaves at least 2 Mg ha^{-1} of residue on the surface. Similarly, in the intermediate and low precipitation zones, farm compliance is achieved in part by maximizing surface residue levels on the fields that are not planted in the fall.

In some situations the high levels of residue retained for erosion control during the winter of the noncritical year can cause difficulty with subsequent tillage operations. The reason for this is that many farmers lack secondary tillage implements or seeding equipment that can operate in high rates of surface residues or high rates of residue that are incorporated in the shallow layers.

Residues also pose disease problems and may reduce yields of some crops, especially with short rotations. For example, on winter wheat several major root rots are favored by surface residues and/or lack of soil disturbance typical of no-till systems (Cook, 1990). In some cases the pathogens responsible for root rot diseases of wheat are carried over in even low amounts of residue during a year of fallow and can reduce yields of early seeded fall crops (Cook, 1992). Surface residues as well as layered buried residues can also favor root diseases of pea (Allmaras et al., 1988). Even with the longer 3-year winter wheat-barley-pea rotation which helps control soilborne pathogens of wheat, no-till barley in winter wheat stubble is still vulnerable to root diseases because these two small grain crops are susceptible to the same root diseases (Cook, 1990). Thus, it is important that with heavier amounts of residues, the management technology needs to consider disease control of crops.

VIII. Residue Management Strategies

The erosion control system applied often allows for rates of erosion on a field that may be excessive during a part of the rotation so long as an offsetting high level of control is being applied at other times which maintains the rotational average for predicted erosion at an acceptable level. However, excessive erosion even at intermittent times will continue to degrade soil quality and cause offsite damage. Ideally, improvements are needed in residue management that will reduce soil loss to acceptable levels on all fields and all parts of the rotation. Farmers also need management options that will simultaneously maximize the benefits of residues for water conservation.

A. Management for Low Residue Producing Crops

Even distribution and maximum retention of surface residues is paramount for maximum erosion control. Pea harvest in particular often results in uneven distribution of crop residues using standard straw spreaders on the combine. An alternative is a harrowing operation after harvest perpendicular or at an angle to the direction of the straw windrow. Maximum retention of surface residues is best accomplished with a no-till drill that simultaneously places seed and fertilizer with little soil disturbance. One such machine is the cross-slot drill (Baker and Saxton, 1988) that shows promise as an ultra surface residue-conserving tool.

With the cross-slot drill farmers could achieve up to 40 or 50 percent cover which is about the maximum to be expected from pea and lentil crops. Another system that is gaining acceptance by farmers combines cloddiness and residues for erosion control and is known as the shank-seed system (Jennings et al., 1990). Fertilizer is placed through chisel shanks which is followed by seeding with double disk openers all in a single operation. With this system farmers can realistically achieve residue levels of 20 to 30 percent following the low residue crops. The shank-seed system merits attention because it has been observed that erosion on thawing soils following a pea or lentil crop is less with minimum tillage seeding and a rough seedbed than with no-till, even with more residues on the surface.

B. Fallow

As with low residue producing crops, maximum retention of residues is the key to having adequate surface cover at planting time. Each tillage operation usually reduces residue cover by 10 to 20% with surface residue farming equipment. With conventional systems using a primary tillage operation in the fall or spring followed by 3 to 5 summer rodweeding operations to control weeds there is little residue cover left by seeding time. Farmer observations indicate that fall tillage of any kind tends to accelerate overwinter residue disappearance. Delaying tillage until spring tends to preserve residues longer. Fall tillage is used to control weeds and to improve infiltration in areas where soil freezing is common.

Improved residue management strategies should consider substituting chemical weed control for fall tillage and use of wider spaced shank tools on chisel implements to improve infiltration in frozen soils. Russian thistle (*Salsola iberica*) is a major weed in the dryland areas. However, stands after wheat harvest which ultimately produce seed are often sparse with individual plants spaced a few to 10 m apart. Newly developed weed-sensing spray technology that targets individual plants offers new opportunities for economical fall weed control with chemicals which currently is cost prohibitive in the dry areas. A further broad spectrum chemical application in early spring to kill weed growth would allow the first fallow tillage operation, which normally would be done in early March, to be delayed until May. With reasonable management in the summer to control weeds, farmers should realistically achieve 30 to 40% residue cover at seeding time.

The other available option is chemical fallow. Experiments have shown that with no-till, surface cover remains relatively high throughout the fallow season though there is a substantial loss in mass of the residue. For example, at Pullman, WA starting with 9 Mg ha^{-1} of wheat straw and 98% surface cover after harvest, the mass loss after 49 weeks was 76% but surface cover was still 91 percent (Stott et al., 1990). Mass loss after 48 weeks was only slightly less (72 percent) for a 1.7 Mg ha^{-1} loading rate. One question that remains is the effectiveness of the surface cover for erosion control after one year of weathering compared with the original residue.

C. Shallow Incorporation of Heavy Residues

Excessive residues following a winter wheat crop can interfere with fall planting in the case of recropping, and with subsequent spring planting and weed control operations. They can also reduce crop growth through abiotic factors and lower soil temperatures in the spring (Elliott et al., 1978). In the intermediate precipitation zone, heavy wheat residues can interfere with fallow operations in the spring and subsequent weed control. Tillage tools such as disks and angled blade chisels can be used to bury 30 to 40 percent of the residue in the shallow layers which retains some of its effectiveness for erosion control and interferes less with tillage operations. A new implement announced recently by DMI, Goodfield, IL[1] features a disk assembly mounted ahead of a chisel plow which offers new opportunities for managing buried/surface residue ratios (Conservation Technology Information Center, 1992). The disk and chisel plow units have independent hydraulic controls which allow the operator without stopping for adjustments to alter mixing and covering residues for different slope positions where rates may be highly variable in the same field.

D. Longer Crop Rotations

Limited rotations such as alternating winter wheat-food legume often do not meet conservation compliance because of the high frequency of the low residue producing crop and difficulties with handling heavy wheat residues. One possibility for increasing erosion protection is to extend the rotation by including a spring cereal such as barley to become a sequence of winter wheat-spring barley or wheat-food legume. This allows for a second winter of cereal stubble which will reduce the rotational average of predicted erosion. A limitation of growing barley is that it could decrease the wheat base of the farm which is not a viable option for many growers.

E. Annual Cropping in Place of Fallow

An option for keeping cover on the land in the dryland areas is annual cropping as a replacement for fallow. Fallow is practiced to increase available water for winter wheat and help stabilize yields at economic levels. However, fallow wastes water and results in overall reduced crop water use efficiency. An alternative that should be considered in the dry areas is a no-till system with winter wheat-spring wheat or continuous spring wheat cropping. With spring wheat it is feasible to control winter annual grass weeds. In the intermediate precipitation zone where fallow is often practiced every third year it may be possible to replace fallow with a water conserving food legume such as lentil or chickpea (*Cicer arietinum*). A no-till planting system would help ensure residue

[1]Trade names and company names are included for the benefit of the reader and do not imply endorsement or preferential treatment of the product by the U.S. Department of Agriculture.

cover at all times of the year which is important for water conservation and wind and water erosion control.

IX. Research Needs

Some priority research needs include the following:

- Develop residue management methods that extend the benefits of winter wheat residues for erosion control from the noncritical year of the rotation to the critical year of the fall seeded crop. With current practices, benefits of residues are lost through the mechanical action of tillage and enhanced decomposition.

- Develop residue management methods for spring-seeded crops that maximize the amount of surface cover for winter wheat planting.

- Determine the effectiveness for erosion control of residues incorporated in the shallow layers of soil for a range of residue rates and of different crops.

- Determine the effectiveness for erosion control of surface residues as they change over time from weathering and decomposition. Does the effectiveness of a given percent cover change after a year of fallow compared with that of the original residues?

- Develop standardized methods for measuring or determining the weight of residues or percent cover. Include improved residue-to-grain conversion ratios and the variations that can be expected based on grain varieties and growing-season conditions. What is the effect of tillage operations on the ratio of surface/buried residues and the resulting effect on erosion and decomposition rates?

- Determine when and where to allow burning of crop residues.

- Develop plant types for spring crops with improved residue characteristics for erosion control and increased resistance to degradation. This is especially needed of food legumes.

- Develop improved practices for conserving residue during fallow. What are the effects of pesticides on the useful life of crop residues? Are there chemical/tillage combinations that can retain surface residues and extend their benefits for erosion control better than can be done today?

References

Allmaras, R.R., J.M. Kraft, and D.E. Miller. 1988. Effects of soil compaction and incorporated residue on root health. *Ann. Rev. Phytopathol.* 26:219-243.

Austin, M.E. 1965. Land resource regions and major land resource areas of the United States. *USDA Agric. Handbook* No. 296.

Baker, C.J. and K.E. Saxton. 1988. The cross-slot conservation-tillage grain drill opener. *ASAE paper* No. 88-1568. American Society Agricultural Engineering, St. Joseph, MI 49085.

Bristow, K.L., G.S. Campbell, R.I. Papendick, and L.F. Elliott. 1986. Simulation of heat and moisture transfer through a surface residue-soil system. *Agric. Forest Meteorol.* 36:193-214.

Conservation Technology Information Center. 1992. Conservation Impact. Vol. 10, No. 3. West Lafayette, IN.

Cook, R.J. 1990. Diseases caused by root-infecting pathogens in dryland agriculture. p. 215-239. In: R.P. Singh, J.F. Parr, and B.A. Stewart (eds.) *Dryland Agriculture - Strategies for Sustainability.* Advances in Soil Science, Vol 13. Springer-Verlag, New York.

Cook, R.J. 1992. Wheat root health management and environmental concern. *Canadian J. Plant Pathol.* 14:76-85.

Elliott, L.F., T.M. McCalla, and A. Waiss, Jr. 1978. Phytotoxicity associated with residue management. p. 131-146. In: W.R. Oschwald (ed.) *Crop Residue Management Systems.* ASA Special Publication No. 31. Am. Soc. Agron., Crop Sci. Soc. Am., and Soil Sci. Soc. Am., Madison, WI.

Gregory, J.M. 1982. Soil cover prediction with various amounts and types of crop residue. *Trans. Am. Soc. Agric. Eng.* 25:1333-1337.

Horner, G.M., A.G. McCall, and F.G. Bell. 1944. Investigations in erosion control and the reclamation of eroded land at the Palouse Conservation Experiment Station, Pullman, WA. 1931-42. *U.S. Department of Agriculture Technical Bulletin* 860.

Horner, G.M., W.A. Starr, and J.K. Patterson. 1957. The Pacific Northwest wheat region. p. 475-481. In: *Yearbook of Agriculture.* U.S. Government Printing Office, Washington, D.C.

Jennings, M.D., B.C. Miller, D.F. Bezdicek, and D. Granatstein. 1990. Sustainability of dryland cropping in the Palouse: A historial view. *J. Soil Water Conserv.* 45:75-80.

McCool, D.K., M. Molnau, R.I. Papendick, and F.L. Brooks. 1977. Erosion research in the dryland grain region of the Pacific Northwest. Recent developments and needs. p. 50-59. In: *Proceedings of a national conference on soil erosion - Soil Erosion: Prediction and control.* Purdue University, West Lafayette, Indiana. May, 1976. Soil Conservation Society of America, Ankeny, Iowa.

McCool, D.K., H. Kok, and R.C. McClellan. 1990. Cover vs mass relationships for Small Grain Residues. *ASAE Paper* No. 90-2040. American Society of Agricultural Engineering, St. Joseph, MI 49085.

Papendick, R.I. and G.S. Campbell. 1974. Wheat-fallow agriculture: Why, how and when? *Proceedings 2nd Regional Wheat Workshop - Tillage and cultural practices for wheat under low rainfall conditions.* Ankara, Turkey. May 1974. The Rockefeller Foundation and CIMMYT.

Papendick, R.I. and D.E. Miller. 1977. Conservation tillage in the Pacific Northwest. *J. Soil Water Conserv.* 32:49-56.

Papendick, R.I., D.K. McCool, and H.A. Krauss. 1983. Soil Conservation: Pacific Northwest. p. 273-290. In: H.E. Dregne and W.O. Willis (eds.) *Dryland Agriculture.* Agronomy Monograph 23. Am. Soc. Agron., Crop Sci. Soc. Am., and Soil Sci. Soc. Am., Madison, WI.

Papendick, R.I. 1987. Tillage and water conservation: experience in the Pacific Northwest. *Soil Use and Management* 3:69-74.

Ramig, R.E. and L.G. Ekin. 1976. Conservation tillage effects on water storage and crop yield in Walla Walla and Ritzville soils. *Oregon Agric. Exp. Stn. Spec. Report* no. 459.

Ramig, R.E., R.R. Allmaras, and R.I. Papendick. 1983. Water conservation: Pacific Northwest. p. 105-124. In: H.E. Dregne and W.O. Willis (eds.) *Dryland Agriculture.* Agronomy Monograph 23. Am. Soc. Agron., Crop Sci. Soc. Am., and Soil Sci. Soc. Am., Madison, WI.

Stott, D.E., H.F. Stroo, L.F. Elliott, R.I. Papendick, and P.W. Unger. 1990. Wheat residue loss from fields under no-till management. *Soil Sci. Soc. Am. J.* 54:92-98.

Thomas, H. and A.M. Grano. 1971. *Agricultural land and water resource base and productivity.* Pacific Northwest. U.S. Dept. of Agric., Economic Res. Service, Natural Resource Economics Division, Corvallis, OR.

U.S. Geological Survey. 1970. Distribution of principal kinds of soils: Orders, sub-orders, and great groups. Map no. 86. In: *National atlas of the United States of America.* U.S. Dept. of Interior, U.S. Geological Survey, U.S. Government Printing Office, Washington, D.C.

Yoder, D.C., J.P. Porter, J.M. Laflen, J.R. Simanton, K.G. Renard, D.K. McCool, and G.R. Foster. 1994. Cover-Management factor (C). Chapter 5. In: *Predicting Soil Erosion by Water: A Guide to Conservation Planning with the Revised Universal Soil Loss Equation.* Agricultural Research Service, U.S. Department of Agriculture. (In Press).

Zuzel, J.F. and J.L. Pikul, Jr. 1987. Infiltration into a seasonally frozen agricultural soil. *J. Soil Water Conserv.* 42:447-450.

Crop Residue Management Strategies
for the Midwest

D.R. Griffith and N.C. Wollenhaupt

I. Introduction

Crop residue management in the Midwest ranges from complete burial to nearly complete soil cover after planting. Chisel plows, which leave partial residue cover, have been used in a few areas for water or wind erosion control since the 1930s. Researchers and innovative farmers began experimenting with planting into undisturbed residue in the 1960s. Tillage and planting equipment is now available to support a wide range of crop residue management goals.

Farmers have changed their tillage practices to achieve different goals. Adoption of reduced tillage systems has allowed growers to farm more acres and improve profits, due to reduced labor, fuel, and equipment investment. Others are changing tillage practices to meet soil erosion control goals in various state

1-56670-003-5/94/$0.00+$.50
© 1994 by CRC Press, Inc.

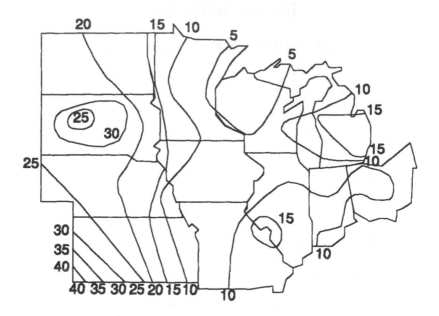

Figure 1. Moisture deficit, cm, between the first spring occurrence of 13°C and the last autumn occurrence of 13°C or first autumn freeze (corn growing season) in the North Central region. (From Aceves-Navarro et al., 1989.)

and federal programs. Strong educational programs and government cost sharing on new farming practices have also contributed to farmer adoption of tillage practices which reduce soil erosion.

The following discussion will focus on tillage and residue management systems in use today and their development in the Midwest region based on local climate, soils, erosion potential and tradition.

II. Midwest Diversity in Climate and Crops

The 12 states composing the Midwest Region provide great diversity in rainfall, temperature, soil topography and physical properties, and in crops grown. From east to west, rainfall deficit during the growing season ranges from 10 cm to 30 cm or more (Figure 1). Average growing degree days for corn range from 2000 on the northern fringe to 4000 on the southern fringe of the region (Figure 2).

Gently rolling dark colored prairie-derived soils dominate the central part of the region, while more sloping, light colored forest derived soils are more common on the southern, northern and eastern fringes of the region. Areas of

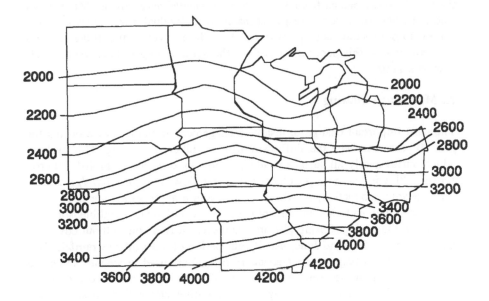

Figure 2. Average growing degree days (GDD) between 25% freeze probability dates. (From U.S. Dept. of Commerce, 1972.)

well drained, coarse-textured soils are found more frequently in the northern part of the region.

Corn and soybean are the dominant crops in the central part of the region, but there is a shift to grain sorghum and wheat in the lower rainfall areas of Kansas, Nebraska, and the Dakotas. Forage crops for beef cattle are more common in northern Missouri, and southern portions of Illinois, Indiana, and Ohio. Growing forages (especially alfalfa) and corn for silage serves the dairy industry in Minnesota, Wisconsin, and Michigan. Vegetable crops are also important in these three states.

This climatic, soil, and crop diversity not only affects soil erodability, but also affects kind and amount of residue remaining after harvest, and crop response to various forms of residue management. Thus, the following discussion will address major residue management issues within the region, but these may not relate to all areas in the Midwest.

III. Major Tillage Systems in Use

Recent surveys summarized by the Conservation Technology Information Center (CTIC) (1991) report significant increases in surface residue after planting during the past 10 years. Crop acreage with surface cover of 30% or more in the CTIC survey ranged from 37% to 53% of planted acres among Midwestern states in 1991. However, acreage estimates for individual tillage systems in the survey do not include acreage with less than 30% cover. Thus, tillage system use estimates in this paper are based on the authors' experience, using CTIC data as a guide.

A. Moldboard Plowing

The most important change in residue management in the Midwest during the past 10 years has been the shift away from moldboard plowing. This system, which buries nearly all surface residue, is now used on less than 40% of crop acreage. Plowing is more popular when corn follows corn or alfalfa/grass sod, for vegetable production, or where heavy manure applications are used.

Depressional soils are usually plowed in the fall while erodible soils are plowed in the spring. Spring plowing, however, leaves soil without cover for most of April, May, and June, usually our period of most intense rainfall. One to three shallow tillage passes (secondary tillage) are used to prepare a seedbed on plowed soil, using some combination of disks, field cultivators or other finishing tools. Harrows or firming baskets are often attached to the same piece of equipment.

B. Chiseling (and other primary tillage)

Chisel plows, large disks, and "sub-soilers" are being used (sometimes in combination) to provide full width tillage to a depth greater than 15 cm. The same secondary tillage used after moldboard plowing is used for seedbed preparation.

These systems provide less inversion of soil and residue than moldboard plowing. Because they leave a rough soil surface, several secondary tillage passes are required before planting. Surface residue after planting may range from almost none (after soybean) to 50% following corn. Tillage systems based on these primary tillage tools can provide adequate water erosion protection following corn, wheat, or sod on slopes of 2 to 4%, but may not provide adequate surface residue on steeper slopes, even following high-residue crops.

These systems are used on 30-35% of cropland acres in the region, but are most popular on the gently rolling prairie soils where corn/soybean rotations dominate.

C. Shallow, Full Width Tillage

Small diameter disks, field cultivators with sweeps and other shallow, full width tillage tools are used in one to three passes, without primary tillage, to prepare a seedbed on about 10% of the crop acreage in the Midwest. Following corn and other high-residue crops, surface cover after planting of 30% to 50% is possible with shallow tillage. However, these systems are most popular following low-residue crops such as soybean or silage corn. In these situations, little surface residue remains after planting with any full-width tillage system.

D. Ridge Planting

A tillage system which includes planting on an elevated ridge with limited soil disturbance was developed in the 1960s. As practiced in the Midwest, ridges are normally formed in corn or soybean fields at last cultivation. No further tillage is done until planting time the following spring. Ridge scrapers, mounted ahead of planter units push old-crop residue and some soil to row middles. Seed is then planted into the residue-free, moist soil cleared by the scrapers. Ridges are again reshaped at last cultivation and the cycle is repeated.

Ridge planting is used on about 5% of the crop acreage in the Midwest, but is more popular (10-12% of crop acreage) in more western states of Nebraska, Minnesota, and South Dakota. East of the Mississippi River, ridge planting is used for one percent, or less, of crop acreage.

Following high-residue crops such as corn, ridge planting is likely to provide adequate erosion control on highly erodible land if some ridge remains in the row area after planting. This directs rainfall to row middles which are protected by residue. Following soybeans, residue cover provided by the ridge system may not be adequate on highly erodible land unless ridges and rows are on the contour.

E. No-Till Planting

Both researchers and innovative farmers began to look at seeding corn into a slot or 1" to 2" tilled strip in the mid 1960s. Fluted coulters were usually used ahead of planter units to cut through crop residue and loosen soil. Small "packer wheels" were used to close seed slots. Many equipment variations are available today for "no-till" planting. The goal remains, however, to leave most of last year's crop residue on the soil surface, while achieving good germination of the crop being planted.

No-till acreage for the Midwestern region is now about 10-12% of crop acreage. This system is most widely used in the southern portions of Ohio, Indiana, and Illinois and in Missouri. However, no-till acreage is rapidly expanding into other portions of the Midwest. Compliance with erosion control provisions of the 1985/1990 farm bills has played a part in no-till expansion, however, much of the new no-till acreage is not on highly erodible land. Apparently, good managers are adopting no-till as a means of reducing costs and increasing net profit.

For many years, no-till planting was used mostly for corn, but no-till soybean has been increasing rapidly in recent years and acreage is now split about evenly between the two crops. No-till is now used on about 70 percent of soybean which is double-cropped after wheat in the southern one-third of the region. Conserving time and moisture are important advantages for these late-planted soybean. Better herbicides for season-long weed control and availability of drills designed for no-till soybean have taken much of the risk from early planted, single-crop no-till soybean. Drilling no-till soybean is currently the fastest growing aspect of no-till planting.

Among the conservation tillage systems now used, no-till leaves the most residue cover; thus the greatest protection from erosion. Surface cover after planting often ranges from 40-60% in soybean residue and 70-90% in corn residue. It is important to remember that surface cover not only depends on type of previous crop, but also on amount of growth and yield, how evenly the residue was distributed at harvest, amount of over-winter residue decay, and amount of residue disturbance by planter attachments and fertilizer applicators.

Over-winter cover crops such as rye, wheat or hairy vetch may be needed in addition to no-till planting to provide adequate protection on highly erodible land following silage corn or low-yielding soybean. While cover crops are now used on less than 10% of the highly erodible land in the region, technology necessary for their successful use is fairly well established.

IV. Soil Erosion vs. Residue Management

The USDA Soil Conservation Service reports that 30% of the cropland acres in Midwestern states have been classified as highly erodible land (HEL) (CTIC, 1991). This same report indicates that only 39% of this land had been adequately treated through 1991, leaving a total of 11,000,000 hectares in the Midwest still in need of protection.

Conservation plans have been signed by farmers in which they have agreed to bring the HEL land into compliance with Food Security Act regulations by 1995. Most of these plans include residue management as the principal means of reducing soil erosion. As stated earlier, no-till and ridge planting systems usually leave enough residue cover to achieve significant protection from erosion. However, the most popular form of conservation tillage in the Midwest is some form of "mulch" tillage, usually including primary tillage with a chisel.

Research from Indiana (Griffith et al., 1986) and Nebraska (Shelton et al., 1990) indicates that tillage systems which include both chiseling and disking often leave less than 30% of the soil residue covered, even in heavy-residue corn stubble. Post-plant fertilizer application and cultivation for weed control would further reduce residue cover and protection from erosion early in the growing season.

The Soil Conservation Service and the Equipment Manufacturers Institute (1992) have provided estimates of the effect on residue cover of a single pass

Table 1. Estimates of residue cover after single chisel plow operation

	Percent residue remaining	
Implement	Non-fragile	Fragile
Chisel plows with:		
Sweeps	70-85	50-60
Straight points	60-80	40-60
Twisted points	50-70	30-40
Combination chisel plows		
Coulter chisel plows with:		
Sweeps	60-80	40-50
Straight points	50-70	30-40
Twisted points	40-60	20-30
Disk chisel plows with:		
Sweeps	60-70	30-50
Straight points	50-60	30-40
Twisted points	30-50	20-30

(Developed by SCS-USDA and the Equipment Manufacturers Institute, 1992.)

with 9 different types of chisels (Table 1). Residue remaining ranges from 20% for a disk chisel with twisted points used in soybean residue to 85% for a standard chisel equipped with sweeps and used in corn residue.

Farmers' use of chisel-based tillage systems to manage residue is reflected in a Wisconsin survey from 16 southern Wisconsin counties in 1991 (Enlow et al., 1992). Chisel plowing was the primary tillage choice for most farmers, with 60% using this system for corn following corn. However, 40% of the chisel plow farmers were not meeting the residue management goal in their conservation plan (usually 30% soil cover after planting), although most farmers assumed that they were practicing conservation tillage.

Table 2 summarizes percent soil cover for various secondary tillage practices after chisel plowing corn stubble in Wisconsin (Wollenhaupt and Reisdorf, 1991). As in other studies, disking following chisel plowing reduced ground cover to below 30%. Replacing the disk with implements equipped with sweeps increased residue on the soil surface, and use of a field cultivator with sweeps after disking left more residue than disking alone.

The studies discussed above show that, when properly used, "mulch" tillage can provide significant protection from erosion. The Wisconsin survey suggests, however, that many farmers in the Midwest have not yet adapted their chisel plows and secondary tillage to maximize surface residue.

Table 2. Residue cover resulting from various kinds of secondary tillage in chiseled corn residues, Wisconsin

Tillage tool	No. of observations	Average cover, %	Made goal in cons. plan, %
No-till planting w/o chiseling	5	70	100
No secondary tillage w/chiseling	6	37	67
Single pass			
Tandem disk	24	27	63
Field cultivator	12	34	100
Soil finisher	18	31	67
Other	7	25	29
Multiple passes			
Tandem disk (2x)	6	23	50
Field cultivator (2x)	5	30	60
Tandem disk + field cultivator	9	32	67
Tandem disk + cultipacker	6	36	83

(From Wollenhaupt and Reisdorf, 1991.)

V. Tillage System Adaptability

Researchers have documented the value of crop residues in controlling erosion from wind or water. Conservationists are promoting crop residue management as a practice that many farmers can use to meet conservation compliance objectives. Farmers, however, must evaluate expected yield, cost, and profitability in choosing tillage systems, in addition to erosion potential. Long-term research and farmer experience have both shown variable response to the range of conservation tillage systems available.

A. Yield Trials

Ohio trials comparing no-till with plowing began in the early 1960s on well drained, sloping, low organic matter Wooster silt loam (Typic Fragiudalf) and on poorly drained, dark Hoytville silty clay loam (Mollic Ochraqualf) (Table 3). No-till had higher yields on the well drained soil in all rotations, but lower yield in continuous corn on the poorly drained soil compared to plowing. Yields were equal or nearly equal in rotation on the poorly drained soil.

Long-term Indiana studies (Griffith et al., 1988) on a dark, poorly drained silty clay loam (fine-silty, mixed mesic Typic Haplaquall) show no-till corn and soybean yields reduced by 13 and 11% following corn, but following soybean, yields were reduced only 3 and 6% compared to plowing (Table 4). Ridge-

Table 3. Corn yield response to tillage and rotation, Ohio

Rotation	Hoytville silty clay loam[a] 1963-1984		Wooster silt loam[b] 1962-1984	
	No-till	Plow	No-till	Plow
	------------------------------Mg ha[-1]------------------------------			
Continuous corn	7.03	7.85	8.16	7.28
Corn-soybeans	8.10	8.10	8.29	7.41
Corn soybeans-meadow	7.97	8.35	9.16	8.41
LSD 0.05	0.025		0.034	

[a]Mollic Ochraqualf; [b]Typic Fragiudalf
(From Dick, et al., 1986a and Dick, et al., 1986b.)

Table 4. Effect of tillage and rotation on corn and soybean yield, 1980-1990, Chalmers silty clay loam (Typic Haplaquoll), Indiana

Previous crop	Tillage	Corn yield	Soybean yield
		----------------Mg ha[-1]----------------	
Corn	Fall plow	10.80a	3.56a
	Fall chisel	10.55a	3.43a
	Ridge-plant	10.67a	3.43a
	No-till	9.42b	3.16b
Soybeans	Fall plow	11.61ab	3.30a
	Fall chisel	11.42bc	3.09a
	Ridge-plant	11.74a	3.16a
	No-till	11.24a	3.09a

Within rotation, yields followed by the same letter are not significantly different at the 0.05 level
(From Griffith et al., 1992a.)

tillage yields were not significantly different from plowed yields for corn or soybeans in either rotation. Yield trials on a one percent organic matter, high silt soil (fine-silty, mixed mesic Typic Ochraqualf) in southern Indiana (Table 5) showed that no-till had best yields among a range of tillage systems for both corn and soybean in rotation.

In Missouri, a long-term continuous corn tillage experiment which includes moldboard plow, chisel, disk and no-till treatments has been underway since 1975 at the Greenley Agronomy Research Center. The study is on Putnam silt loam (fine, montmorillonitic, mesic Mollic Albaqualf) and Mexico silt loam (fine, montmorillonitic, mesic Udollic Ochraqualf), commonly called claypan (poorly drained) soils. Corn grain yields were not significantly affected by

Table 5. Effect of five tillage systems on corn growth and yield on Clermont silt loam (Typic Ochraqualf), 1983-1986, Indiana

	Continuous corn		Corn after soybean	
Tillage system	Height, 8 wks. cm	Yield Mg ha⁻¹	Height, 8 wks. cm	Yield Mg ha⁻¹
Spring plow	165a[a]	8.22b	165b	8.22c
Fall chisel	170ab	8.60ab	160b	8.16c
Spring disk	163b	8.29b	157b	8.60b
Ridge-plant	170ab	8.22b	170b	8.79b
No-till	180a	8.85a	193a	9.35a

[a]Data followed by same letter are not significantly different at 0.05 level
(From Griffith et al., 1988.)

Table 6. Corn grain yields as influenced by tillage on Putnam (Mollic Albaqualf) and Mexico (Udollic Ochraqualf) silt loam soils, 10 yr. average, Missouri

	------------------------------Tillage system------------------------------				
	Plow	Chisel	Disk	No-till	Average
	--------------------------------Mg ha⁻¹------------------------------				
High yield	10.28	10.28	10.97	10.66	10.53
Low yield	0.88	1.19	1.19	0.69	1.00
10 yr mean	5.89	6.02	6.08	5.83	5.96

(From Wollenhaupt and Buchholz, 1986.)

tillage (Table 6) from 1975-1986 (Wollenhaupt and Buchholz, 1986), a period which included years with a wide range in yield potential.

An Iowa study (Chase and Duffy, 1991) reports corn and soybean yields from 1978-1987 from a Floyd soil (fine loamy, mixed, mesic Aquic Hapludoll). No-till corn yields were lower than those for moldboard plowing in continuous corn, but the two systems provided similar yields for corn and soybean grown in rotation (Table 7). Also in Iowa, Brown et al. (1989) reported no-till corn yields to be less than yields with reduced tillage (disk or field cultivate) and fall moldboard plowing on a poorly drained Taintor soil (Typic Argiaquall) and a somewhat poorly drained Mahaska soil (Aquic Argiudoll). Eight year average soybean yields were not significantly affected by tillage.

No-till continuous corn yields in Minnesota were reduced by 15% on a poorly drained soil (Typic Albaqualf) and by 5% on a moderately well drained soil (Typic Argiudoll). Chisel and moldboard plowing produced equal yields on both soils (Table 8).

Table 7. Corn and soybean yield response to tillage on a poorly-drained Floyd soil (Aquic Hopludoll) in Iowa

Crop rotation	Tillage system	Yield (Mg ha[-1])
Continuous corn	Moldboard plow	8.54a[a]
	Chisel plow	8.22b
	Ridge-till	7.97bc
	No-till	7.85c
Corn after soybean	Moldboard plow	9.16a
	Chisel plow	9.16ab
	Ridge-till	8.91ab
	No-till	8.91b
Soybean after corn	Moldboard plow	2.69a
	Chisel plow	2.69a
	Ridge-till	2.55b
	No-till	2.62a

[a]Yields followed by same letter are not significantly different (P= 0>05). (From Chase and Duffy, 1991.)

Table 8. Effect of tillage and soil drainage class on continuous corn yield, Minnesota

Tillage system	Somewhat poorly drained[a] 1987-1990	Moderately well drained[b] 1985-1989
	-------------------------Mg ha[-1]-------------------------	
Moldboard	9.42	10.36
Chisel plow	9.49	10.17
No-till	8.03	9.85

[a]Ames fine sandy loam, Typic Albaqualf; [b]Tama silt loam, Typic Argiudoll. (From Moncrief et al., 1990 and Moncrief et al., 1991.)

In a Wisconsin study which evaluated solid seeded soybean response to cultivar and tillage, no-till soybean yielded about 135 kg ha[-1] less than soybean with moldboard plowing (Philbrook et al., 1991). In a separate study, Oplinger and Philbrook (1992) found that no-till seeding rates needed to be approximately 15% higher than those for moldboard plow tillage if equivalent yields were to be expected. In these studies, cultivar had the same relative yield independent of tillage system, where brown stem rot (Phialophora gregata) was not present. This disease has been found to be more serious with no-till planting (Meese et al., 1991).

Table 9. Effect of tillage on the yield of grain sorghum in a wheat-sorghum-fallow system, 1978-1987

Tillage system	Yield (Mg ha^{-1})
3 or 4 tillage operations[a]	3.92
Reduced[a]	4.12
No-till	4.22
LSD 0.05	0.215

[a]All tillage with V-blade or rod weeder.
(From Norwood, 1988.)

The effects of tillage on continuous corn production have been variable in Wisconsin. Al-Darby and Lowery (1986) found that ridge-plant, chisel-disk, and no-till systems produced yields equal to corn with moldboard plowing on Griswold silt loam (fine-loamy, mixed, mesic Typic Argiudoll) and Plainfield loamy sand (sandy, mixed, mesic Typic Udipasment). In a corn hybrid performance trial, Carter and Barnett (1987) found no-till yields to be 92 to 95% of moldboard plow tillage. They attributed lower no-till grain yields to delayed seasonal development, especially with full season hybrids in 1984, a cool spring year.

Bundy et al. (1992) found, in a 3 year Wisconsin study, that corn yields were equivalent between four tillage systems including no-till and moldboard plow on a Manawa silty clay loam (fine, mixed, mesic Aquollic Hapludalf), but, no-till yields were significantly lower than yields with plowing on a Plano silt loam. On both soils, no-till corn had the greatest yield increase with added nitrogen, suggesting higher fertilizer N requirements in no-till. Research in the central and southern portions of the Midwest has shown that properly applied N is used efficiently by both no-till and plowed corn (Mengel, 1992). When corn and soybean were grown in rotation, moldboard plow and no-till yields were similar (Meese et al., 1991).

Hesterman et al. (1988) report in a Michigan study that no-till and moldboard plow system corn yields were not significantly different on a somewhat poorly drained Capac loam (fine loamy, mixed, mesic Aeric Ochraqualf) and a well drained Oshtemo sandy loam (coarse loamy, mixed, mesic Typic Hapludalf).

In a long-term study in Kansas, (Norwood, 1988) on the drier western fringe of our Midwest Region, grain sorghum yields (Table 9) were 8% higher with no-till planting than with the conventional 3 or 4 full-width tillage operations.

B. Things That Influence Tillage System Adaptability

Farmers who are thinking about changing tillage and/or planting systems to leave more crop residue on the soil surface also want to maintain or increase productivity and profits. From tillage/crop production research and farmer experience, several factors can be identified which might influence the adaptability of no-till, ridge-till, and other reduced tillage systems.

Most studies which report lower yields for no-till or other high residue systems, cite delayed crop growth and development, and reduced plant populations as most likely causes. Cool soil, allelopathy and improper seed placement are often related to the problems. The following factors are likely to influence these problems and reduced tillage success.

** Soil drainage and texture -- No-till planting seems especially well adapted on well drained soils, often providing greater yield potential than full-width tillage systems. Ridge planting usually provides yields equal to fall plowing on dark, poorly drained soil where no-till is often at a disadvantage. On soils with intermediate drainage, there has tended to be little difference in yield potential among tillage systems. Reduced no-till yields on dark, poorly drained soil may be compounded by soil compaction, tillage pans, or low fertility. Improved soil physical properties with time from surface residue tends to favor no-till over other systems on low and medium organic matter soils.

** Ground Cover -- Soils warm in response to radiant energy from the sun. Crop residues which intercept or reflect this energy, effectively insulate the soil, leading to cooler, wetter soil and slower plant development. Crops such as corn for grain, wheat and sod produce the most ground cover, but soybean may also limit warming where residues are concentrated and not evenly spread on the soil.

** Latitude -- The shorter growing season and cooler soil temperature at planting time in the northern one-half of the region tends to put no-till planting at a disadvantage.

** Crop rotation -- No-till and other surface residue systems tend to be more competitive with clean tillage when crops are rotated. This is especially true for corn, since the heavy surface residue in no-till mono-culture may cause slow growth, both from cool soil and from allelopathy. Soybean are usually planted after corn, allowing more time for soil warming.

Of course, crop management decisions also influence conservation tillage success. Proper fertilization, weed, insect, disease, and rodent control, and equipment adjustment are all necessary, along with the farm operator's ability to properly manage these inputs.

C. Tillage Adaptability to Specific Locations

Choosing the "right" tillage system for a particular field is sometimes not a simple process. The influence of soil, climate, rotation, and management experience on success with different tillage systems should be considered. A final decision will most likely be made based on potential for net profit and

D.R. Griffith and N.C. Wollenhaupt

Table 10. Yield coefficients for tillage systems, Indiana

	Soil tillage-management group[a]								
	1	2	3	4	5	6	7	8	9
Rotation corn									
Fall plow	1.00	1.00	1.00	1.00	1.00	1.00	1.00	1.00	1.00
Fall chisel	1.00	1.00	1.03	1.03	1.00	1.00	1.00	1.00	1.00
Spring plow	0.96	1.00	1.00	1.05	0.98	0.98	1.00	1.00	1.03
Spr. disk/fld. cult.	1.00	1.00	1.03	1.05	1.00	1.00	1.03	1.03	1.03
Ridge	1.00	1.00	1.06	1.03	0.98	1.00	1.03	1.06	1.03
No-till	0.98	1.06	1.06	0.98	0.96	0.98	1.06	1.10	1.10
Continuous corn									
Fall plow	0.93	0.93	0.93	0.93	0.93	0.93	0.93	0.93	0.93
Fall chisel	0.91	0.93	0.98	0.96	0.93	0.93	0.96	0.98	0.98
Spring plow	0.87	0.93	0.93	0.98	0.91	0.91	0.95	0.95	0.95
Spr. disk/fld. cult.	0.89	0.93	0.98	0.98	0.91	0.93	0.96	0.98	0.98
Ridge	0.91	0.91	0.98	0.98	0.89	0.93	0.96	1.00	1.00
No-till	0.85	0.95	0.92	0.90	0.87	0.91	1.00	1.03	1.05
Rotation soybean									
Fall plow	1.00	1.00	1.00	1.00	1.00	1.00	1.00	1.00	1.00
Fall chisel	0.96	1.00	1.05	1.03	1.00	1.00	1.03	1.05	1.05
Spring plow	0.96	1.00	1.00	1.05	1.00	1.00	1.00	1.00	1.02
Spr. disk/fld. cult.	0.96	1.05	1.05	1.05	1.00	1.00	1.03	1.05	1.05
Ridge	0.96	1.05	1.05	1.05	0.98	1.00	1.03	1.05	1.05
No-till	0.92	1.07	0.99	0.96	0.97	0.98	1.05	1.10	1.12

[a]Soil group descriptions and example soil series:

1. Dark, poorly drained silty clay loams to clays, 0 to 2% slope. Examples: Brookston, Chalmers, Pewano.

2. Light (very low organic matter), somewhat poorly drained silt loams, nearly level to gently sloping, overlying very slowly permeable fragipan-like pans that restrict plant rooting and water movement. Examples: Cleremont, Avonburg.

3. Dark, poorly drained, "high water table," loamy sands and sandy loams on 0 to 2% slope. Because of moderately coarse and coarse surface textures, these soils are subject to severe wind erosion and damage to young plants by blowing sand. Examples: Maumee, Lyles, Granby.

4. Dark muck soils with greater than 30% organic matter, poorly drained, nearly level, subject to severe wind erosion if left unprotected. Examples: Adrian, Carlisle, Edwards, Haughton.

5. Light (low organic matter), somewhat poorly drained silt loams to silty clay loams with high clay subsoils, nearly level to 4% slopes. Example: Blount.

6. Same as 5, but without high clay subsoil. Examples: Crosby, Fincastle.

Table 10. -- continued

7. Light (low organic matter), well and moderately well drained upland soils with silt loam to sandy loam surface texture and high clay subsoil on slopes of 2 to 6 % (subject to moderate water erosion). Example: Morley.

8. Same as 7, but without high clay subsoil. Also, coarse textured terrace soils subject to moderate wind erosion and/or drought. Examples: Miami, Fox.

9. Light (low organic matter), well drained soils on slopes greater than 6 % that are subject to very severe water or wind erosion. Examples: Miami, Russel, Alford. In addition, Sands. Example: Plainfield.

(From Doster et al., 1991.)

erosion control, and on eligibility for government programs. Relating tillage system to yield potential, however, should be a first step in choosing a system.

To aid in tillage system decisions, several states in the Midwest have classified soils into tillage-management groups, both for corn and for soybeans. In Ohio, (Triplett et al., 1973) named soil series were placed into five groups according to soil properties and their influence on no-till planting. In Indiana, over 400 soil series have been placed into tillage-management groups for continuous corn, rotation corn, and rotation soybeans. Both adaptability ratings (Steinhardt et al., 1990) and yield coefficients (Doster et al., 1991) were assigned to tillage systems, ranging from fall plowing to no-till, for each soil group. Soil characteristics considered in grouping soils included drainage, texture, organic matter, and slope. The coefficients (Table 10) estimate percent yield increase or decrease for conservation tillage systems compared to fall moldboard plowing in crop rotation.

Such tillage system ratings are a good guide to farmers in these states who are considering major changes in tillage. For regions with differing soils and climate, however, these tillage system evaluations may not be appropriate. Techniques used in classifying soils and evaluating tillage system adaptability could be used for other areas.

VI. Modified Systems – Strip Preparation

Currently, there is great interest among farmers in the Midwest in modifying no-till and ridge-till systems to increase their areas of adaptability. However, since the modifications alter residue cover, they may affect erosion control in addition to crop growth.

Many farmers are preparing a strip 15 to 10 cm wide for each row in otherwise untilled fields. Benefits which may result from this strip preparation are warmer and drier soil in the seed zone, reduced allelopathy (especially when

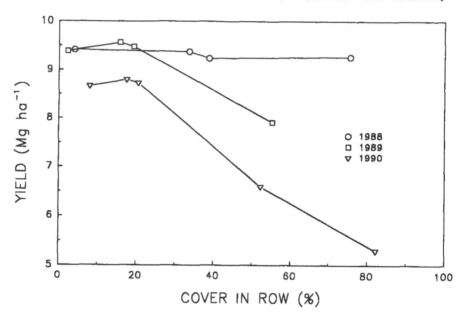

Figure 3. Effect of residue-free band width on mean grain yield, Ames, IA, 1983-85. (From Kaspar et al., 1990.)

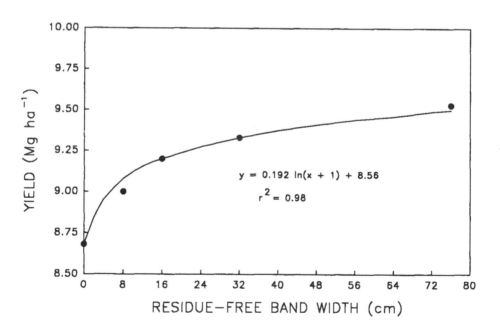

Figure 4. Effect of cover in the row on corn yields. (From Moncrief et al., 1991.)

Table 11. Continuous corn response to tillage, 4-yr means from four soil types, Indiana

Tillage	4 wk. stand, % of seed rate	Soil temp., first 8 wk.[a]	Plant ht. at 8 wk.	Grain yield
	%	C	cm	Mg ha^{-1}
Moldboard plow	86	23.6	155	7.31
Strip rotary[b]	84	21.7	141	6.90
No-till	87	20.9	132	6.60

[a]Daily maximum taken in row; [b]Tilled strips were 25 cm wide × 10 cm deep.
(From Griffith et al., 1973.)

Table 12. Continuous corn response to tillage, Ontario, Canada, sandy loam soil, 1982-1984

Tillage	Plant population	Plant ht., June	Grain yield
	Plants ha^{-1}	cm	Mg ha^{-1}
Moldboard plow	57,700	67.3	9.37
Rotovator strip[a]	57,400	60.9	8.86
Cultivator strip[a]	57,500	56.5	8.66
No-till	50,000	53.6	7.60

[a]Tilled strips were 25 cm wide × 10 cm deep.
(From Vyn and Raimbault, 1992.)

Table 13. Continuous corn response to strip preparation on a Typic Haplaquoll soil, Lafayette, Indiana, 1992

Tillage	5 wk. mean daily soil temp.	Leaves at 4 wk.	Leaves at 8 wk.	Time to tassel
	C°	number	number	days
No-till	16.7	2.6	7.2	82.7
Spoked wheels	17.1	3.0	7.6	81.3
3 fluted coulters	17.8	2.8	7.3	81.3
P.U. mini-tiller[a]	17.3	2.8	7.6	82.0
Moldboard plow	18.7	3.4	8.3	78.0

[a]A steel blade was used to loosen soil strip 12.5 cm wide × 12.5 cm deep.
(From Griffith et al., 1992b.)

corn follows corn, wheat or rye), reduced soil density in the seed zone, more uniform seed placement, and more vigorous early rooting.

Both residue removal and/or zone tillage are being used alone or in combination. Double disks or spiked wheels are usually used to move residue from the row and a series of non-powered, fluted coulters are most often used to disturb soil in an 18-20 cm band ahead of planter units. Other means that could be used to loosen a band of soil include powered rotary tillage, cultivator sweeps, angled blades to lift soil, and anhydrous ammonia applicator knives with "wings" attached.

A. Research Results

In a Central Iowa study (Kaspar et al., 1990), removing residue from a band only 8 cm wide for corn significantly increased plant height, decreased days to 50% emergence and days to 50% tassel, decreased grain moisture at harvest, reduced barrenness, and increased grain yield by 0.3 Mg ha^{-1} (Figure 3). Farther north in the Midwest Region, Minnesota researchers (Moncrief et al., 1991) found that residue removal from the row area was essential for success of no-till corn in 1989 and 1990, but not in the drier spring of 1988 (Figure 4). They concluded that greater than 20% in-row surface cover was likely to lower yield potential. Nelson and Shinners (1988), in Wisconsin, determined that a 32 cm wide residue free band provided soil temperature that produced corn emergence on the same day as with bare ground tillage. They also noted that residue may not stay in place once it is removed, and that width of planter gauge wheels should be considered in choosing strip width.

Few research studies have evaluated strip tillage. Both Indiana (Griffith et al., 1973) and Ontario (Vyn and Raimbault, 1992) studies (Tables 11 and 12) found that rotary tilled strips about 19 cm wide improved growth, maturity and yield of corn compared to standard no-till planting, but these parameters were still significantly reduced when compared to full width tillage. Preliminary Indiana results (Griffith et al., 1992b) (Table 13) using non-powered fluted coulters and soil lifters show some increase in soil temperature and plant maturity for these treatments but they did not provide plant growth equal to that with plowing.

There is considerable speculation that use of strip preparation for each row in the fall might provide a warmer seedbed at spring planting and avoid the wet soils often found under heavy residue at planting. Results from fall use of strip preparation are not yet documented, however.

B. Strip Preparation and Erosion

Moving residue from the row and/or loosening soil can cause serious water erosion problems if rows run up and down slope. In heavy residue, strip preparation may leave enough surface cover to meet standards of the Food Security Acts of 1985/1990, but still leave fields subject to serious erosion. On highly erodible land, strip preparation should be used on the contour, or soil loosening equipment which leaves surface cover should be used.

VII. Current Restraints to Adopting Surface Residue Systems

While surface residue tillage systems are increasing in the Midwest, about 60 % of the land classified as highly erodible is still left without adequate protection. The following appear to be significant factors in restricting the adoption of surface residue systems.

** Perception of lower yield -- A bad experience with residue systems in a local area is long remembered, even though lower yields may have been due to poor management or on poorly-drained non-erosive soils.

** Economic weed control -- While some farmers are able to switch to residue systems with little or no added cost for weed control, most will use different herbicides than they used with clean tillage, with some increased cost. Use of new application techniques such as early pre-plant application, directed post-emergence application, and low-rate technology may provide better weed control at reduced cost in residue systems.

** Peculiar problems -- Some problems seldom found with clean tillage influence a few fields planted into residue each year. These include trees after several years of no-till, common stalk borer or stink bug damage after cover crops or weedy residue, rodent damage when planting into stand-over meadow, and increased susceptibility to frost in heavy residue. Use of proper management can overcome some, but not all of these problems.

** Perception of "conservation tillage" -- Many farmers who do not moldboard plow believe that they are practicing soil conservation, even though their surface residue levels are less than 30 %.

** Cash available -- Some farmers do not have cash available and are not able to borrow needed money to buy a no-till drill or equipment to modify their older planters.

** Management ability -- Most people agree that switching to a surface residue system requires some additional management skills, usually in equipment adjustment or in weed or insect control. Many farmers who lack these skills are small operators who farm part-time. These farmers are also more likely to have highly erodible land.

** Lack of knowledgeable professionals -- Both in industry (farm equipment, fertilizer and chemical dealers) and in government agencies (Soil Conservation Service, Cooperative Extension Service), there are not enough people trained as both conservation and crop production professionals. These people who advise farmers one-on-one should have working knowledge and experience with no-till, ridge-till and reduced tillage systems.

** Lack of long-term research -- Many areas still do not have field scale research dealing with adaptation of various tillage/planting systems to unique soil, climatic, and cropping conditions.

Most of the above-listed constraints can be at least partly addressed through education. Transferring known technology to prospective surface residue farmers should be a high-priority challenge for the 1990s.

References

Aceves-Navarro, L.A., R.E. Nield, and K.G. Hubbard. 1989. *Agroclimatic normals for corn in the North-central Region.* North-central Regional Research Publication, Institute of Agriculture and Natural Resources, Univ. of Nebraska, Lincoln, NE.

Al-Darby, A.M. and B. Lowery. 1986. Evaluation of corn growth and productivity with three conservation tillage systems. *Agron. J.* 78:901-907.

Brown, H.J., R.M. Cruse, and T. Colvin. 1989. Tillage system effects on crop growth and production costs for a corn-soybean rotation. *J. Prod. Agric.* 2:273-279.

Bundy, L.G., T.W. Andraski, and T.C. Daniel. 1992. Placement and timing of nitrogen fertilizers for conventional and conservation tillage corn production. *J. Prod. Agric.* 5:214-221.

Carter, P.R. and K.H. Barnett. 1987. Corn-hybrid performance under conventional and no-tillage systems after thinning. *Agron. J.* 79:919-926.

Chase, C.A. and M.D. Duffy. 1991. An economic analysis of the Nashua tillage study: 1978-1987. *J. Prod. Agric.* 4:91-98.

Conservation Technology Information Center (CTIC). 1991. *National survey of conservation tillage practices.* 1220 Potter Dr., West Lafayette, IN.

Dick, W.A., D.M. Van Doren, G.B. Triplett, and J.E. Henry. 1986a. *Influence of long-term tillage and crop rotation combinations on crop yields and selected soil parameters, I. Results obtained for a Mollic Ochraqualf soil.* Research Bulletin 1180, Ohio Agricultural Research and Development Center, Wooster, Ohio.

Dick, W.A., D.M. Van Doren, G.B. Triplett, and J.E. Henry. 1986b. *Influence of long-term tillage and crop rotation combinations on crop yields and selected soil parameters, II. Results obtained for a Typic Fragiudalf soil.* Research Bulletin 1181, Ohio Agricultural Research and Development Center, Wooster, Ohio.

Doster, D.H., S.D. Parsons, D.R. Griffith, G.C. Steinhardt, D.B. Mengel, R.L. Nielsen, and E.P. Christmas. 1991. *Influence of production practices on yield estimates for corn, soybean and wheat.* ID-152, Purdue Univ. Coop. Ext. Ser., West Lafayette, IN.

Enlow, J., N. Wollenhaupt, and T. Reisdorf. 1992. Are we on schedule with conservation compliance? *Proceedings 1992 Fertilizer, Aglime and Pest Management Conference*. Holiday Inn, Middleton, WI. January 21-23. University of Wisconsin, Madison, WI.

Griffith, D.R., E.J. Kladivko, J.V. Mannering, T.D. West, and S.D. Parsons. 1988. Long-term tillage and rotation effects on corn growth and yield on high and low organic matter, poorly drained soils. *Agron. J.* 80:599-605.

Griffith D.R., J.V. Mannering, and J.E. Box. 1986. Soil and moisture management with reduced tillage. In: M. A. Sprague and G. B. Triplett (eds.) *No-tillage and Surface Tillage Agriculture*. Wiley and Sons, New York, NY.

Griffith, D.R., J.V. Mannering, H.M. Galloway, S.D. Parsons, and C.B. Richey. 1973. Effect of eight tillage-planting systems on soil temperature, percent stand, plant growth, and yield of corn on five Indiana soils. *Agron. J.* 65:321-326.

Griffith, D.R., J.F. Moncrief, D.J. Eckert, J.B. Swan, and D.D. Brietbach. 1992a. Influence of soil, climate, and residue on crop response to tillage systems. In: *Conservation Tillage System and Management*. Mid-west Plan Service, Iowa State Univ., Ames, IA.

Griffith, D. R., T. D. West, G. C. Steinhardt, P. Hill, and S. D. Parsons. 1992b. Strip preparation for no-till corn and soybean. CES Paper No. 258, Purdue Univ. Coop. Ext. Ser., W. Lafayette, IN.

Hesterman, O.B., F.J. Pierce, and E.C. Rossman. 1988. Performance of commercial corn hybrids under conventional and no-tillage systems. *J. Prod. Agric.* 1:202-206.

Kaspar, T.C., D.C. Erbach, and R.M. Cruse. 1990. Corn response to seed-row residue removal. *Soil Sci. Soc. Am. J.* 54:1112-1117.

Meese, B.G., P.R. Carter, E.S. Oplinger, and J.W. Pendleton. 1991. Corn/soybean rotation effect as influenced by tillage, nitrogen, and hybrid/cultivar. *J. Prod. Agric.* 4:74-80.

Mengel, D.B. 1992. Fertilizing corn grown using conservation tillage. AY-268, Purdue Univ. Coop. Ext. Ser., W. Lafayette, IN.

Moncrief, J.F., S.D. Grosland, and J.J. Kuznia. 1991. Tillage effects on corn growth, stand establishment, and yield, Isanti County, MN. In: *A Report on Field Research in Soils*. Minnesota Agr. Exp. Sta., Univ. of Minnesota, St. Paul, MN.

Moncrief, J.F., T.L. Wagar, and J.J. Kuznia. 1990. Tillage system and cultivation effects on corn growth and yield on a well-drained silt loam soil. In: *A Report on Field Research in Soils*. Minnesota Agr. Exp. Sta., Univ. of Minnesota, St. Paul, MN.

Nelson, W.S. and K.J. Shinners. 1988. Effects of residue free bands on plant and soil parameters. ASAE Paper No. 88-1039. ASAE St. Joseph, MI.

Norwood, C. 1988. Conservation tillage cropping systems; Southwest Kansas Experiment Station. p. 1-9. In: *Conservation Tillage Research*. Kansas State Univ., AES Report of Progress 542.

Oplinger, E.S. and B.D. Philbrook. 1992. Soybean planting date, row width, and seeding rate response in three tillage systems. *J. Prod. Agric.* 5:94-99.

Philbrook, B.D., E.S. Oplinger, and B.E. Freed. 1991. Solid-seeded soybean cultivar response in three tillage systems. *J. Prod. Agric.* 4:86-91.

Shelton, D.P., E.C. Dickey, P.J. Jasa, and S.S. Krotz. 1990. Tillage system influences on corn residue cover. ASAE Paper No. 90-2041. ASAE, St. Joseph, MI.

Soil Conservation Service, USDA, and the Equipment Manufacturers Institute. 1992. Estimates of residue cover remaining after single operation of selected tillage machines. SCS-USDA, Washington, D.C.

Steinhardt, G.C., D.R. Griffith, and J.V. Mannering. 1990. Adaptability of tillage-planting systems to Indiana soils. AY-210, Purdue Univ. Coop. Ext. Ser., West Lafayette, IN.

Triplett, G.B., D.M. Van Doren, and S.W. Bone. 1973. An evaluation of Ohio soils in relation to no-till corn production. Res. Bul. 1068, Ohio Agr. Res. and Development Center, Wooster, OH.

U.S. Dept. of Commerce. 1972. Minnesota Weekly Weather and Crop Bulletin. National Oceanic and Atmospheric Administration, Environmental Data Service, Vol. 59, No. 15.

Vyn, T.J. and B.A. Raimbault. 1992. Evaluation of strip tillage systems for corn production in Ontario. *Soil Tillage Research* 23:163-176.

Wollenhaupt, N.C. and D.D. Buchholz. 1986. Long term tillage study for corn production. 1986. p. 37-40. In: Field Day Report. Greenley Memorial Center, College of Agr., Univ. of Missouri, Columbia, MO.

Wollenhaupt, N. C. and T. Reisdorf. 1991. Tillage equipment usage and crop residue management by Wisconsin farmers. In: *Keeping Current*, Vol. 1, No. 7. Environmental Resources Center, University of Wisconsin, Madison WI. 4 pp.

Residue Management Strategies – Great Plains

P.W. Unger

I. Introduction

The Great Plains is a vast subhumid to semiarid, North American mid-continent region extending from Texas into the Canadian prairie provinces. In the USA, it covers about one-fifth of the total area of the contiguous states. Precipitation in the Great Plains has an east-west gradient, and annual amounts range from about 750 mm at the eastern boundary (at about 98° west longitude) to about 300 mm at the Rocky Mountains that form the western boundary. Because precipitation is limited, dryland crop production in the Great Plains is strongly influenced by the amount of water stored in soil. Besides being limited, precipitation is also highly variable among and within years, which results in highly variable crop yields from year to year. Irrigation, which could stabilize production, is limited to portions of the Great Plains where sufficient surface or ground water is available. However, most crops are grown without irrigation; hence, to stabilize production, development of improved water conservation practices has received much attention throughout the Great Plains.

To increase the amount of water in soil at crop planting time, fallowing has been widely used in the Great Plains since the latter part of the 19th century

(Haas et al., 1974). Fallowing is the practice wherein no crop is grown during the season when a crop might normally be grown. Proponents emphasize the water-conserving, weed-controlling, and crop-yield-stabilizing virtues of fallowing while critics emphasize the low water-storage efficiencies and wind- and water-erosion problems associated with the practice. Despite the controversy, fallowing still is widely used in much of the Great Plains.

Early explorers called the Great Plains the Great American Desert (Webb, 1931) because precipitation was limited, there were few springs or streams, and the landscape was relatively flat and treeless. These conditions still exist, but knowledgeable people today realize that the Great Plains is not a desert. There have been periods, however, when the Great Plains appeared to be a desert. One of those was during the 1930s when a major drought plagued the region. The effects of the drought were intensified by crop production practices being used by farmers during that period. The practices were largely those that settlers from the eastern USA and from Europe brought with them when they settled the region in the latter decades of the 1800s.

In developing the land for crops (mainly small grains), farmers plowed under the native grasses. After crop harvest, cattle were allowed to graze the residues. For subsequent crops, clean tillage that involved total incorporation of the remaining crop residues became the common seedbed preparation practice. Crop production involving clean tillage was satisfactory in years of above-average precipitation. But crops failed and the bare soil was ravaged by the unrelenting winds that prevailed during the drought. Many forecasters thought that the region had no future (Stewart, 1990).

A major consequence of the 1930s' drought was the recognition of the importance of crop residues on the soil surface for controlling erosion, which led to the development of the stubble-mulch farming system. Use of this system permits satisfactory crop production under conditions where adequate residues are retained on the soil surface to control erosion. Stubble mulch tillage is a type of conservation tillage, provided adequate residues are retained on the soil surface to meet the requirements of the conservation tillage definition.

II. Early Residue Management Research

The stubble-mulch tillage system was developed primarily to control wind erosion, but its value for controlling water erosion was soon recognized (McCalla and Army, 1961). Stubble mulch tillage undercuts the surface at a depth of 7 to 10 cm with a sweep or blade implement to control weeds and prepare a seedbed, yet retains most residues on the soil surface to help control erosion. The degree of erosion control achieved by surface residues is influenced by the amount present, surface coverage provided, and residue anchorage (resistance to movement) in the soil. For controlling wind erosion, surface residues reduce wind speeds at the soil-air interface to below the threshold level for soil particle movement. Water erosion is reduced because surface residues

Table 1. Water storage, runoff, and evaporation from field plots at Lincoln, Nebraska, 10 April to 27 September 1939

Treatment	Storage (mm)	Runoff (mm)	Evap. (mm)	Evap. loss (%)[a]
Straw, 2.2 Mg ha⁻¹, normal subtillage	30	26	265	83
Straw, 4.5 Mg ha⁻¹, normal subtillage	29	10	282	88
Straw, 4.5 Mg ha⁻¹, extra loose subtillage	54	5	262	82
Straw, 9.0 Mg ha⁻¹, normal subtillage	87	Trace	234	73
Straw, 17.9 Mg ha⁻¹, no tillage	139	0	182	57
Straw, 4.5 Mg ha⁻¹, disked in	27	28	266	83
No straw, disked	7	60	254	79
Contour basin listing	34	0	287	89

[a] Based on total precipitation, which was 321 mm for the period.
(Adapted from Russel, 1939.)

intercept the energy of falling raindrops, thus reducing soil aggregate dispersion and surface sealing. By maintaining greater infiltration rates, soil transport across the surface is reduced because runoff is reduced.

"Trash and crop residues on the surface check runoff and allow more of the water to be absorbed by the soil." These words by Hallsted and Mathews (1936) indicate that early researchers in the semiarid Great Plains also recognized the value of surface residues for conserving soil water. A few years later, Russel (1939) illustrated the effects of different residue amounts and management practices on soil water conservation (Table 1).

Although retaining residues on the surface helped control erosion by wind and water and increased water storage, crop production under high-residue conditions was not practical at that time because herbicides were not yet available for controlling weeds and suitable equipment was not available for preparing the seedbed, planting the crop, and cultivating the soil. Since the early results and observations, extensive research has been conducted throughout the Great Plains regarding the effects of stubble mulch tillage and other residue-retaining practices on controlling erosion, conserving water, and soil physical and chemical properties.

Table 2. Straw mulch effects on soil water storage efficiency at Sidney, Montana; Akron, Colorado; and North Platte, Nebraska, 1962 to 1969

Mulch rate (Mg ha^{-1})	Fallow period precipitation (mm)	Water storage efficiency (mm)
0	355	16
1.7	355-549	19-26
3.4	355-648	22-30
6.7	355-648	28-33
10.1	648	34

(From Greb et al., 1967.)

III. Soil Water Conservation

A. Infiltration and Evaporation

Crop residues influence soil water conservation through their effects on water infiltration and on soil water evaporation. However, the contributions of residues to increased infiltration and reduced evaporation often are difficult to separate. Hence, the overall effects of residues on soil water conservation or contents generally were reported.

Soil water storage during fallow progressively increased in the northern and central Great Plains as the amount of wheat (*Triticum aestivum* L.) straw on the soil surface was increased from 0 to 10 Mg ha^{-1} (Table 2). For that study (Greb et al., 1967), water storage efficiency during fallow ranged from 16% with no residues to 34% with the high mulch rate. In the southern Great Plains, storage efficiency during fallow after winter wheat was 23% with no residues and 46% with 12 Mg ha^{-1} of wheat straw mulch (Table 3). Dryland grain sorghum [*Sorghum bicolor* (L.) Moench] grown after fallow yielded 1.78 Mg ha^{-1} with no mulch and 3.99 Mg ha^{-1} with the high mulch rate (Unger, 1978).

Field studies at numerous locations in the Great Plains have shown that soil water storage and/or crop yields have been increased by retaining crop residues on the soil surface by use of reduced- or no-tillage practices. Some selected references are Black and Power (1965), Doran et al. (1984), Good and Smika (1978), Lavake and Wiese (1979), Norwood et al. (1990), Phillips (1964, 1969), Power et al. (1986), Smika and Wicks (1968), Tanaka (1989), Unger (1984a), Unger and Wiese (1979), Unger et al. (1971), Wicks and Smika (1973), and Wilhelm et al. (1986, 1989).

Water storage and crop yield responses to surface residues varied among locations and due to residue types and amounts. Water storage generally increased with increased amounts of residues present. For example, Wilhelm et al. (1986) developed the following regression equations that relate surface residues to soil water storage at planting:

Table 3. Straw mulch effects on soil water storage during fallow,[a] water use efficiency, dryland grain sorghum yield, total water use, and water use efficiency at Bushland, Texas, 1973-1976

Mulch rate (Mg ha⁻¹)	Water storage[b] (mm)	Storage efficiency[b] (%)	Grain yield (Mg ha⁻¹)	Total water use (mm)	WUE[c] (kg m⁻³)
0	72 c[d]	22.6 c	1.78 c	320	0.56
1	99b	31.1 b	2.41 b	330	0.73
2	100b	31.4 b	2.60 b	353	0.74
4	116 b	36.5 b	2.98 b	357	0.84
8	139 a	43.7 a	3.68 a	365	1.01
12	147 a	46.2 a	3.99 a	347	1.15

[a]Fallow duration of 10 to 11 months.
[b]Water storage determined to 1.8-m depth. Precipitation averaged 318 mm.
[c]Water use efficiency based on grain produced, growing precipitation, and soil water changes.
[d]Column values followed by the same letter are not significantly different at the 5% level (Duncan's multiple range test).
(From Unger, 1978.)

Maize: $Y = 174 + 6X$ ($r^2 = 0.84$)
Soybean: $Y = 175 + 8X$ ($r^2 = 0.71$)

where Y is available soil water at planting (mm) and X is the amount of surface residues applied after harvest of the previous crop (Mg ha⁻¹). Other equations developed by Wilhelm et al. (1986) related crop yields to surface residue amounts. These were:

Maize: $Y = 2.91 + 0.13X$ ($r^2 = 0.80$)
Soybean: $Y = 1.53 + 0.09X$ ($r^2 = 0.84$)

where Y is grain yield (Mg ha⁻¹) and X is the amount of crop residue (Mg ha⁻¹) left on the surface after harvest of the previous crop. Thus, each Mg ha⁻¹ of residues on the surface increased maize yields an average of 0.13 Mg ha⁻¹ and soybean yields an average of 0.09 Mg ha⁻¹.

The above favorable responses occurred where surface residue amounts were relatively large and 50% or more of the residues were retained on the surface through use of reduced- or no-tillage practices. Where initial residue amounts were low or weed control was poor, water conservation and/or yields with surface residue-retaining tillage often were lower than or equal to those with conventional tillage (Army et al., 1961; Phillips, 1969; Wiese and Army, 1958, 1960; Wiese et al., 1967). Poor yield responses to low residue amounts resulted from greater runoff and evaporation.

Gerard (1987) measured runoff of natural rainfall on Miles fine sandy loam (fine-loamy, mixed, thermic Udic Paleustalf) as affected by surface residue management treatments (Table 4). As compared with bare soil, runoff was

Table 4. Runoff losses and surface cover in spring of 1986 on a miles fine sandy loam soil

Treatement	Runoff from natural rainfall[a] (mm)	(%)	Surface cover (%)
Check	73	36.8	0
7.5 Mg ha⁻¹ straw incorporated	11	5.6	4
15.0 Mg ha⁻¹ straw incorporated	35	17.7	14
7.5 Mg ha⁻¹ straw conservation-tilled[b]	9	4.3	56
15.0 Mg ha⁻¹ straw conservation-tilled[b]	5	2.4	96
Ryegrass	24	12.2	88

[a]Total recorded rainfall 19.9 cm.
[b]Conservation-tilled consisted of incorporating one-half of the straw with the soil and leaving the other half of the straw on top of the soil
(Adapted from Gerard, 1987.)

reduced where straw was incorporated and further reduced where the use of conservation tillage retained most of the straw on the surface. Where residue amounts were low, which is common with dryland crops in the Great Plains, runoff was greater from no-tillage than from stubble mulch tillage watersheds (Jones et al., 1991). Although runoff was greater with no-tillage, soil water contents at planting of winter wheat and grain sorghum were or tended to be greater with no-tillage on Pullman clay loam (fine, mixed, thermic Torrertic Paleustoll) at Bushland, Texas (Figures 1 and 2), apparently because soil water evaporation was greater with stubble mulch tillage. With stubble mulch tillage, the soil often becomes air dry to the depth of soil disturbance by tillage, and this dry soil must be rewet before any water storage can occur at greater soil depths.

Smika (1976) evaluated surface residue effects on soil water evaporation during a 34-day period following 13.5 mm of rainfall. No additional rain fell during the period. One day after the rain, water contents were similar with no-tillage (herbicides only), minimum-tillage (combination of stubble mulch tillage and herbicides), and conventional-tillage (stubble mulch) treatments (Figure 3A). Water contents were greatest with no-tillage and least with conventional tillage after 34 days (Figure 3B). Surface residue amounts during the study were 1.2 Mg ha⁻¹ with conventional-, 2.2 with minimum-, and 2.7 with no-tillage treatments.

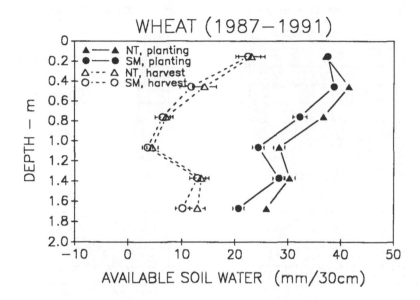

Figure 1. Tillage system effects on soil water content at planting and harvest of dryland winter wheat, Bushland, Texas. (Adapted from Jones, 1992.)

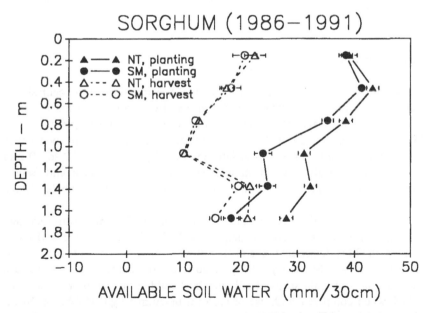

Figure 2. Tillage system effects on soil water content at planting and harvest of dryland grain sorghum, Bushland, Texas. (Adapted from Jones, 1992.)

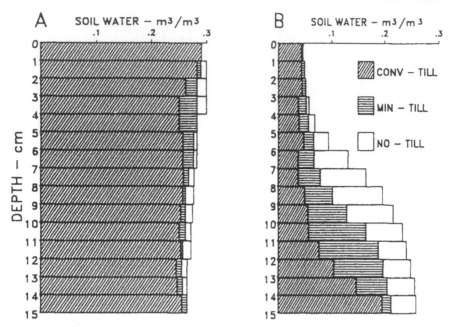

Figure 3. Soil water content to a 15-cm depth 1 day (A) and 34 days (B) after 13.5 mm rainfall as influenced by tillage treatments (CONV-TILL, conventional tillage; MIN-TILL, minimum tillage; NO-TILL, no-tillage). (From Smika and Unger, 1986.)

B. Snow Trapping

Snow comprises a major portion of the annual precipitation in the central and northern Great Plains. Therefore, snow trapping by crop residues has received much attention for water conservation purposes is those regions (Willis et al., 1983). Soil water storage at the 0.3- to 0.9-m depth during three overwinter periods averaged 5 to 54 mm with fall moldboard plowing and 32 to 90 mm with no-tillage (standing wheat stubble) in North Dakota (Black and Bauer, 1985). Water storage was intermediate with disking and V-blade tillage treatments.

Where standing stubble is present, snow trapping and, hence, water storage increases with increases in stubble height (Table 5). The increases in soil water content at the May determination closely followed the water contents of the snow in April (Black and Siddoway, 1977). In addition to stubble height per se, use of alternating 5-m-wide strips of tall and short stubble increased water storage by 30% when compared with use of uniform medium-height stubble in North Dakota (Willis and Frank, 1975). Increases in snow trapping and water storage also can be achieved by maintaining strips of tall grasses (Aase et al., 1976; Black and Aase, 1988; Black and Siddoway, 1976). For maximum effectiveness,

Table 5. Snow depth, water equivalent of snow, and total soil water content as affected by stubble management

Date	Stubble height, mm			
	0[a]	150	280	380
		Snow depth, mm		
2 April 1975	61	119	198	259
		Water equivalent of snow, mm		
2 April 1975	13	25	43	56
		Total soil water content, mm		
12 May 1975	267	277	287	307

[a]This treatment involved conventional disk tillage.
(Adapted from Black and Siddoway, 1977.)

stubble and grass barrier orientation should be perpendicular to prevailing winter winds. From a management viewpoint, the distance between strips should be equivalent to one or more widths of the equipment being used.

Through the years, tillage practices have greatly changed in the Great Plains, and these have greatly improved soil water storage and crop yields. Greb (1979) gave an example of this for the period from 1916 to 1990 (estimated) for winter wheat production in a wheat-fallow system at Akron, Colorado (Table 6). Initially, moldboard plowing plus harrowing to create a dust mulch was the common practice in the Great Plains. Tillage systems that retain crop residues on the surface are now used in many cases. From the first period (1916-30) to the last period (1975-90), water storage during fallow was almost doubled and grain yield was more than doubled. Besides improved water conservation due to surface residues, improved varieties, fertility practices, and weed control undoubtedly contributed to the yield increases. Greater infiltration, lower evaporation, and improved snow trapping with increasing surface residues contributed to the improvements in water conservation with changing tillage practices.

IV. Erosion Control

Soil erosion due to wind and water occurs in all Great Plains states (Table 7). Based on the totals, erosion in excess of T (estimated soil loss tolerance under existing management, according to 1982 National Resources Inventory) is more prevalent due to wind than to water in all states, except Nebraska and Wyoming (USDA, 1989). For those portions of most states that lie within the Great Plains

Table 6. Progress in fallow systems with respect to water storage and wheat yields at Akron, Colorado

| Years | Tillage during fallow[a] | Fallow water storage | | Wheat yield |
		(mm)	(% of prec.)	(Mg ha^{-1})
1916-30	Maximum tillage; plow, harrow (dust mulch)	102	19	1.07
1931-45	Conventional tillage; shallow disk, rod weeder	118	24	1.16
1946-60	Improved conventional tillage; begin stubble mulch in 1957	137	27	1.73
1961-75	Stubble mulch; begin minimum tillage with herbicides in 1969	157	33	2.16
1975-90	Projected estimate; minimum tillage; begin no-tillage in 1983	183	40	2.69

[a]Based on 14-month fallow, from mid-July to second mid-September.
(Adapted from Greb, 1979.)

(west of about 98° west longitude), the potential for wind erosion generally is much greater than for water erosion because precipitation decreases from east to west.

A system to predict the influence of climate, soil factors, crops grown, crop residue management practices, etc. on water erosion has been developed (Laflen et al., 1991a, 1991b). A wind erosion prediction system is under development also (Hagan, 1991). Both systems will aid in understanding erosion problems and processes, and in developing improved erosion control strategies. The system for wind erosion should be of major value for the Great Plains where most wind erosion occurs.

A. Wind Erosion

Wind erosion in the northern Great Plains causes most damage during February, March, and April when overwinter precipitation and freezing and thawing have reduced soil aggregates to erodible sizes, and where soils recently tilled for spring crops have not yet been covered by vegetation. In the central Great Plains, the potential is greatest in February and March because winter wheat does not adequately cover the soil at that time (Allmaras, 1983). The potential for wind erosion where winter wheat is grown in the southern Great Plains is similar to that of the central Great Plains. In all regions, wind erosion can be severe where summer crops are grown that produce only limited amounts of

Table 7. Cropland areas where erosion is in excess of T in the Great Plains states

Region and states	Cropland (1000 ha)	Erosion in excess of T			
		-----Water-----		-------Wind------	
		(1000 ha)	(%)	(1000 ha)	(%)
Northern Plains					
Kansas	11,739.6	1,811.3	15	2,019.7	17
Nebraska	8,205.9	1,870.0	23	591.5	7
North Dakota	10,942.7	1,043.8	10	2,533.9	23
South Dakota	6,858.5	967.9	14	1,178.2	17
Region total	37,746.7	5,693.0	15	6,323.3	17
Southern Plains					
Oklahoma	4,681.6	564.8	12	971.2	21
Texas	13,484.4	1,982.4	15	6,280.4	47
Region total	18,166.0	2,547.2	14	7,251.6	40
Mountain					
Colorado	4,290.9	434.0	10	2,188.5	51
Montana	6,959.5	693.7	10	3,371.4	48
New Mexico	976.4	47.8	5	264.5	27
Wyoming	1,047.1	83.9	8	39.1	4
Region total	13,273.9	1,259.4	9	5,863.5	44
Overall totals	69,186.6	9,499.6	14	19,438.4	28

(Adapted from USDA, 1989.)

residue. A major problem area is the large sandy soil area of the southern Great Plains where cotton (*Gossypium hirsutum* L.) is the primary crop.

The principles of wind erosion have been extensively studied, and Lyles et al. (1983) gave a thorough review of the subject. The basic principles of wind erosion control are:

1. Establish and maintain vegetation or vegetative cover. (Surface cover.)
2. Produce or bring nonerodible aggregates to the soil surface. (Nonerodible surface aggregates.)
3. Reduce field width along the direction of prevailing erosive winds. (Field width.)
4. Roughen the land surface. (Surface roughness.)

These principles remain constant, but control practices to achieve them vary in space and time as new cropping and management systems are developed. Crop residues have a major direct impact on the first principle, and may indirectly influence the remaining principles. The following discussion stresses the effects of surface cover provided by crop residues for controlling wind erosion.

Table 8. Effect of tillage machines on surface residue remaining after each operation

Tillage machine	Approximate residue maintained (%)
Subsurface cultivators	
Wide-blade cultivator and rodweeder	90
Mixing-type cultivators	
Heavy-duty cultivator, chisel, and other type machines	75
Mixing and inverting disk machines	
One-way flexible disk harrow, one-way disk, tandem disk, offset disk	50
Inverting machine	
Moldboard and included disk plow	10

(Adapted from Anderson, 1968.)

1. Surface Cover

Wind erosion is effectively controlled when soil surfaces are completely covered by growing crops or crop residues. As surface cover decreases, the potential for erosion increases. According to the CTIC (1990) definition, residue covers equivalent to at least 1.1 Mg ha^{-1} of small grain residues are needed to protect soils during critical wind erosion periods. The residues must be anchored in soil to avoid being blown away.

The approximate percentages of residues retained on soils after each operation by various tillage implements are given in Table 8. In Table 9, actual percentages of soybean and corn residues remaining after planting are given for various tillage systems. Most residues are retained on the surface when a no-tillage system is used. With no-tillage, residue disappearance results primarily from decomposition; some may be blown away. Of the tillage implements considered, most residues are retained on the surface when subsurface cultivators equipped with sweeps, blades, and/or rodweeders are used. These implements are used for stubble mulch tillage, which is widely used for dryland grain crops throughout much of the Great Plains. Stubble mulch tillage is now considered to be conventional tillage for these crops in many cases in the Great Plains.

Because stubble mulch tillage undercuts the surface, it is an effective method for retaining surface residues and, hence, for controlling wind erosion. Of course, adequate residues must be available. A disadvantage of long-term use of stubble mulch tillage is the development of a shear plane or compacted zone at the primary tillage depth in some cases, which may require occasional deeper-than-normal tillage or tillage with a different implement for alleviation of the problem.

Table 9. Residue cover remaining after planting

Tillage and planting operation	Residue cover remaining, %	
	Soybean[a]	Corn[b]
No-till plant	62.3	56.0
Blade plow, plant	31.7	41.4
Field cultivate, plant	23.6	32.7
Blade plow, till-plant	22.9	19.8
Till-plant	18.5	30.1
Disk, plant	17.8	28.2
Disk, field cultivate, plant	16.1	19.6
Chisel plow, disk, plant	13.2	19.1
Disk, disk, plant	13.0	17.6

[a]Residue was from two soybean varieties and two row spacings, which yielded an average of 2.6 Mg ha^{-1}.
[b]Residue was averaged for 2 years. Grain yields were 8.1 Mg ha^{-1} in Year 1 and 6.2 Mg ha^{-1} in Year 2.
(Adapted from Dickey et al., 1990.)

Good residue retention is achieved also by using chisel-type implements, but weed control is more difficult with chisels than with sweep or blade implements. Disk implements can be used where large amounts of residues are present or where weed populations warrant such drastic weed control measure. However, disks must be used with caution to avoid destroying too much of the residues. Where most residues have been incorporated with a disk or other implements, a possible means of returning some residues to the surface would be to use a chisel or sweep implement as a second operation. Such operation may result in returning adequate residues to the surface to provide satisfactory wind erosion control. Moldboard plowing eliminates most residues from the surface.

The effects of residue type (wheat or sorghum), quantity, and orientation on soil loss from a sand loam under wind tunnel conditions are given in Table 10. Standing residues had a distinct advantage over flat residues, and wheat residues had a distinct advantage over sorghum residues for reducing erosion. With about 4.5 Mg ha^{-1} of wheat residues, erosion was virtually eliminated, whereas about three times that amount was needed to obtain similar control with sorghum residues (Finkel, 1986).

Wind erosion is a major problem where cotton is grown on the sandy soils of the southern Great Plains because, after harvest, only the stalks remain, which have limited value for erosion control purposes. However, systems are in use or being developed that help control erosion in this region. These systems normally include growing crops that provide a wind barrier or a surface cover during the critical wind erosion period. Bilbro and Fryrear (1988) grew sorghum and millet [*Pennisetum americanum* (L.) K. Schum] cultivars as wind barriers in the unplanted rows of skip-row cotton. Cotton lint yield was not significantly

Table 10. Average effects of kind and orientation of crop residue on erosion of a sandy loam soil by wind of uniform velocity

Quantity of crop residue above soil surface	Quantity of soil eroded in a wind tunnel			
	Covered with wheat residue		Covered with sorghum residue	
	Standing 25 cm high	Flat	Standing 25 cm high	Flat
	----------------------------Mg ha[-1]--------------------------------------			
0	35.8	35.8	35.8	35.8
0.56	6.3	19.0	29.1	32.5
1.12	0.2	5.6	18.1	23.3
2.24	T[a]	0.2	8.7	11.9
3.36	T	T	3.1	4.9
6.72	T	T	T	0.4

[a]T = trace, insignificant.
(Adapted from Finkel, 1986.)

reduced due to competition by the barrier crops, and the various barrier crops were equally effective per unit height for reducing wind velocity, which reduced wind erosion. On Amarillo fine sandy loam (fine-loamy, mixed, thermic Aridic Paleustalf), which has an annual soil loss potential of over 100 Mg ha[-1], tilling the soil and applying cotton gin trash to the surface reduced soil loss to below the assumed annual tolerance level of 11 Mg ha[-1] (Fryrear and Skidmore, 1985). Ivey (1992) planted cotton by no- or reduced-tillage methods into winter wheat, which was used as a cover crop. The wheat was killed in early spring, and the remaining residues provided excellent protection against wind erosion.

2. Nonerodible Surface Aggregates and Surface Roughness

Nonerodible aggregates at the soil surface and extent of soil surface roughness are influenced by such factors as soil texture, tillage implement used, depth and speed of tillage, and soil water content at the time of tillage. Residue management practices should have little immediate influence on erodibility of surface aggregates, but do influence aggregate stability if given practices are used over an extended period of time (Skidmore et al., 1986; Unger, 1984b). Residue management practices influence surface roughness by affecting the degree of soil reconsolidation that occurs after tillage. Residue incorporation may actually result in a rougher soil surface per se than, for example, stubble mulch tillage. The overall roughness of the surface, namely, that resulting from the combined influence of the soil itself plus surface residues, should increase with increasing amounts of residues retained on the surface. Increases in this overall roughness decreases wind movement at the soil-air interface, which decreases the potential for wind erosion.

3. Field Width

With respect to wind erosion, field width pertains to the distance in a field between barriers or objects that interfere with the erosion process along the path of erosive winds. Typical barriers or objects that influence field width include growing crops, crop residues, plowed soil having a rough surface of nonerodible material, grass strips, and row of trees. For stripcropping, which is widely used in the central and northern Great Plains, field strips are alternately cropped or fallowed, with the strips lying perpendicular to the prevailing erosive winds. In successive years, the previously-cropped strip is fallowed, and vice versa. Narrow strips (6 to 30 m wide) are needed on sandy soils to control wind erosion, whereas strips over 90 m wide can be used on finer-textured soils (Black and Bauer, 1983; Siddoway and Fenster, 1983). Using stubble mulch or no-tillage methods, which retain most crop residues on the soil surface, is important to improve wind erosion control where stripcropping is used to reduce field width.

B. Water Erosion

The principles for controlling water erosion are closely related to those for water conservation, namely, minimizing runoff and maintaining favorable conditions for water infiltration. Hence, those runoff control and infiltration maintenance practices discussed for water conservation are appropriate also for water erosion control in most cases.

The single most effective practice for reducing water erosion is to maintain a surface cover in contact with soil. The cover may be a growing crop or crop residues. A 90% cover of the soil surface by crop residues reduces erosion by 93% as compared with that from bare soil (Wischmeier and Smith, 1978). The surface residues reduce erosion by reducing the energy of raindrop impact on the surface, thus reducing aggregate dispersion and surface sealing that could reduce water infiltration (increase runoff), and by retarding the runoff rate, thus providing more time for water infiltration and for transportable soil particles to settle from the water.

The effects of percent surface cover resulting from different tillage and planting operations involving different crops on soil losses are given in Table 11. In all cases, surface cover was lowest and soil loss was greatest for systems involving moldboard plowing; no-tillage planting gave opposite results. Except for soybean residues, no-tillage planting provided at least 30% surface cover after planting and good erosion control. However, even with soybean residues for which surface cover was 27%, soil loss was only 11.4 Mg ha^{-1} with no-

Table 11. Measured surface cover and soil loss for various tillage and planting systems

Residue type, slope & soil texture Tillage and planting operations	Residue cover	Cumulative soil loss	Erosion reduction from moldboard
	(%)	(Mg ha⁻¹)	(%)
Corn residue, 10% slope, silt loam[a]			
Moldboard plow, disk, disk, plant	7	17.5	--
Chisel plow, disk, plant	35	4.7	74
Disk, disk, plant	21	4.9	72
Rotary-till, plant	27	4.3	76
Till-plant	34	2.5	86
No-till plant	39	1.6	92
Soybean residue, 5% slope, silty clay loam[a]			
Moldboard plow, disk, disk, plant	2	32.0	--
Chisel plow, disk, plant	7	21.5	32
Disk, plant	8	23.7	26
Field cultivate, plant	18	17.0	46
No-till plant	27	11.4	64
Wheat residue, 4% slope, silt loam[b]			
Moldboard plow, harrow, rod weed, plant	9	9.4	--
Blade plow 3 times, rod weed, plant	29	2.7	72
No-till drill	86	4.5	96
Oat residue, 10% slope, silt loam[c]			
Moldboard , disk, harrow, plant	4	56.0	--
Disk, disk, harrow, plant	5	42.6	24
Blade plow, disk, harrow, plant	10	47.0	16
No-till plant	39	11.2	80

[a] 51 mm water in 45 minutes.
[b] 70 mm water in 75 minutes.
[c] 64 mm water in 60 minutes.
(Adapted from Dickey et al., 1990.)

tillage; it was 32.0 Mg ha⁻¹ with moldboard plowing, which resulted in only 2% cover under the same crop and soil conditions.

On Pullman clay loam with a slope of < 1.0% at Bushland, Texas, soil loss from a sweep-tillage watershed was over twice the amount as that from a no-tillage watershed, but average amounts for 1984 to 1991 were only about 0.19

Table 12. Sediment losses during dry or wet runs of simulated rainfall on Pullman clay loam as affected by previous crop and tillage method, Bushland, Texas

| | Tillage treatments[a] | | | | | |
	Plow	Roto	Sweep-RR	Sweep-RL	Notill-RR	Notill-RL
	----------------------------Mg ha^{-1}----------------------------					
Wheat						
Dry run	4.8a	2.5b	0.21b	0.26b	0.14b	0.09b
Wet run	4.3a	1.9bc	0.16bc	0.21b	0.12bc	0.09c
Sorghum						
Dry run	4.4a	3.1ab	0.29ab	0.18bc	0.13c	0.10c
Wet run	2.7a	2.1ab	0.24a	0.14ab	0.11ab	0.06b

[a] Tillage treatments were: Plow -- moldboard plowing plus disking, Roto -- rotary tillage, Sweep-RR -- sweep tillage with residues removed, Sweep-RL -- sweep tillage with residues left in place, Notill-RR -- no-tillage with residues removed, and Notill-RL -- no-tillage with residues left in place. (Adapted from Unger, 1992.)

Mg ha^{-1} with no-tillage and 0.40 Mg ha^{-1} with sweep tillage (Jones, 1992). Soil loss was less with no-tillage, even though runoff was greater. Runoff for 1984 to 1991 averaged 43 mm yr^{-1} from the no-tillage and 27 mm yr^{-1} from the sweep tillage watershed. Most runoff occurred during the fallow-after-sorghum phase of the dryland wheat-sorghum-fallow rotation used on the watersheds. Greater runoff during that phase resulted from the sparse surface cover provided by the dryland sorghum in some years.

At Woodward, Oklahoma, two adjacent grassland watersheds were plowed in 1979, then farmed by conventional or conservation tillage methods for wheat grain (1980-84) or graze-out wheat forage (1985-86) production. Sediment yields from 1980 to 1984 on the wheat watersheds were low and similar to that on adjacent range watersheds. Major rainstorms occurred in 1985 and 1986, and annual sediment losses averaged 68 Mg ha^{-1} on the conventional tillage, 3 Mg ha^{-1} on the conservation (no-) tillage, and 0.3 Mg ha^{-1} on the range watershed (Berg et al., 1988). For the period from 1982 to 1986, sediment losses averaged 15.9 Mg ha^{-1} on the conventional tillage, 0.9 Mg ha^{-1} on the no-tillage, and 0.15 Mg ha^{-1} on the range watersheds (Smith et al., 1991). The dominant soils on the watersheds were Quinlan and Woodward loams (Typic Ustochrepts).

For a simulated rainfall infiltration study on Pullman clay loam with <1.0% slope at Bushland, soil losses for dry and wet runs after dryland wheat and grain sorghum were greater with moldboard plowing than with no-tillage (Table 12). Soil losses ranged from 2.7 to 4.8 Mg ha^{-1} on moldboard plowed plots, which was over four times greater than on no-tillage plots for which residues were retained on the surface. To obtain final infiltration rates, water was applied for a longer time to moldboard than to no-tillage treatment plots, and this could have contributed to greater soil losses for the moldboard treatment. However, surface cover provided by residues undoubtedly also was a factor. Surface cover was 2.9% with moldboard plowing and 9.6% with no-tillage after sorghum, and 9.9% with moldboard plowing and 68.4% with no-tillage after wheat (Unger, 1992).

V. Soil Organic Matter

A major consequence of converting grassland to cropland in the Great Plains was a sharp decline in soil organic matter (OM) concentrations. Haas et al. (1957) studied the effect of cropping on soil N and carbon (C) changes at 14 Great Plains locations for which the period of cropping ranged from 30 to 43 years. The N loss ranged from 24 to 60%, and averaged 39% over a 36-year period for all locations. Percent organic C losses were similar, but slightly greater than the N losses.

The rates of loss of soil N and C were strongly influenced by tillage and cropping systems used and to a lesser extent by crops grown (Bauer and Black, 1981; Haas et al., 1957; Hobbs and Brown, 1957, 1965; Johnson et al., 1974; Unger, 1968, 1972, 1982). In general, N and C decreased more rapidly with residue-incorporating tillage than with stubble mulch tillage that retained residue on the surface, with cropping systems involving fallow than where crops were grown annually, and with row crops than with drilled small grain crops. Applications of manure were highly effective for avoiding major decreases in soil N and C concentrations. Use of rotations that included green manure crops also reduced the decline rates for N and C.

The decline in soil N and C was most rapid when the land was first developed for crop production. With time, the decline rate slowed and there was a trend toward new steady-state levels. The declines, however, resulted in soil structural deterioration in many cases, which aggravated water conservation problems through soil aggregate dispersion, surface sealing, and decreased water infiltration. Reduced biological activity in soil due to lower organic substrate materials contributed to soil structure deterioration (Power et al., 1984).

Increased concerns for the environment (water quality, air quality, global warming, greenhouse effect, etc.) and increased emphasis on sustaining soil productivity have resulted in major interests in OM concentrations and transformations in recent years. Soils have the potential for either increasing or decreasing the OM concentration in various parts of the environment, depending on how they or, more precisely, how crop residues are managed.

No- and reduced-tillage systems retain greater amounts of crop residues on or near the soil surface and, therefore, help maintain soil OM concentrations at greater levels than residue-incorporating tillage or residue-removal practices (Follett and Peterson, 1988; Havlin et al., 1990; Havlin and Schlegel, 1990; Hooker et al., 1982; Lamb et al., 1985; Unger, 1991; Wood et al., 1990).

While both types of tillage may return essentially the same amounts of crop residues to the soil, no-tillage practices result in greater OM concentrations because tillage that hastens OM decline is avoided. No-tillage was especially effective for maintaining OM and N at relatively high levels at or near the soil surface (Follett and Peterson, 1988; Unger, 1991), which is of major importance for stabilizing soil aggregates, thus increasing the potential for more rapid water infiltration and improved soil water conservation. Improved water conservation has improved crop yields, as previously discussed, which, in turn, return more residues to the system, thus providing the potential for even greater OM concentrations in the soil.

VI. Soil pH

A potential consequence of using tillage practices that retain crop residues on the soil surface is the decline of soil pH to levels that may adversely affect crop production. The effect may be direct or through rapid deactivation of certain herbicides used in the crop rotation (Wicks et al., 1988). The potential is greatest where nitrogen fertilizers are surface-applied and not mixed with soil, as where no-tillage is practiced. Near North Platte, Nebraska, pH declined with both the tillage and no-tillage treatments during an 18-year study, but the decline during all phases of the rotation (fallow, wheat stubble, or sorghum) was always greater with the no-tillage treatment (Wicks et al., 1988). No definite pH trends were found after 6 or 8 years of no-tillage in a wheat-sorghum-fallow rotation at Bushland, Texas (Unger, 1991). However, in that study, nitrogen fertilizer was not applied.

VII. Summary

The Great Plains is a vast mid-continent region that extends from Texas into the Canadian prairie provinces and from about 98° west longitude at the east to the Rocky Mountains at the west. The region is subhumid to semiarid, having mean annual precipitation of about 750 mm at the east and 300 mm at the west. Originally, native grasses covered the region. In the latter decades of the 1800s and early decades of the 1900s, much of the region was converted to cropland for small grains, using mostly the farming practices the settlers brought with them from the eastern states and from Europe. Initially, when precipitation was favorable, crop yields were favorable. However, the farming practices (clean tillage) being used contributed to severe wind erosion problems during the 1930s when a major drought plagued the region. A major outcome of that drought was the development of the stubble mulch tillage system. Stubble mulch tillage undercuts the soil surface to control weeds and prepare a seedbed, yet retains crop residues on the surface for controlling erosion. Besides controlling wind erosion, stubble mulch tillage also is beneficial for aiding water erosion control and for water conservation purposes. Stubble mulch tillage is now classified as a conservation tillage method, provided adequate crop residues are available to provide adequate surface cover to protect soils against erosion.

The stubble mulch tillage system now is considered the conventional tillage method in many parts of the Great Plains where small grain crops predominate. Since the 1950s and 1960s, crop production systems involving herbicides for weed control have received much attention in the Great Plains. The herbicides may be used in no-tillage systems or in herbicide-tillage combination systems. Crop yields usually are as good or better with these reduced tillage systems as with conventional tillage.

A major limitation to effective erosion control and greater water conservation with reduced tillage (residue management) systems is the low amount of residues produced by dryland crops in some cases. Where residue amounts are inadequate, surface residues alone may not control erosion or enhance water conservation. In such cases, some tillage may be needed to roughen the surface or to provide depressions for capturing potential runoff water. An especially critical region is the sandy soil region of the southern Great Plains where the potential for wind erosion is great and where cotton is the dominant crop. Cotton residues have limited effectiveness for controlling erosion; however, use of barrier or cover crops is showing promise as an effective erosion control practice.

Crop residue management is an important component of most cropping systems throughout the Great Plains. Besides controlling erosion and enhancing water conservation, crop residue management has a major influence on soil physical, chemical, and biological properties. Reduced tillage practices are helping to maintain soil organic matter at greater levels, thus helping to maintain soil productivity and impacting water and air quality, and helping to preserve the environment.

References

Aase, J.K., F.H. Siddoway, and A.L. Black. 1976. Perennial grass barriers for wind erosion control, snow management, and crop production. pp. 69-78. In: R. Tinus (ed.) *Shelterbelts on the Great Plains*. Great Plains Agric. Counc. Publ. 78, Univ. Nebraska, Lincoln.

Allmaras, R.R. 1983. Soil conservation: Using climate, soils, topography, and adapted crops information to select conserving practices. In: H.E. Dregne and W.O. Willis (eds.) *Dryland Agriculture*. Agronomy 23:139-153.

Anderson, D.T. 1968. Field equipment needs in conservation tillage methods. pp. 83-91. In: *Conservation Tillage in the Great Plains*. Proc. of a Workshop, Lincoln, Nebraska, February 1968. Great Plains Agric. Counc. Publ. 32, Univ. Nebraska, Lincoln.

Army, T.J., A.F. Wiese, and R.J. Hanks. 1961. Effect of tillage and chemical weed control practice on soil moisture losses during the fallow period. *Soil Sci. Soc. Am. Proc.* 25:410-413.

Bauer, A. and A.L. Black. 1981. Carbon, nitrogen, and bulk density comparisons in two cropland tillage systems after 25 years and in virgin grassland. *Soil Sci. Soc. Am. J.* 45:1166-1170.

Berg, W.A., S.J. Smith, and G.A. Coleman. 1988. Management effects on runoff, soil, and nutrient losses from highly erodible soils in the Southern Plains. *J. Soil Water Conserv.* 43:407-410.

Bilbro, J.D. and D.W. Fryrear. 1988. Wind erosion control with annual plants. pp. 89-90. In: P.W. Unger, T.V. Sneed, W.R. Jordan, and R. Jensen (eds.) *Challenges in Dryland Agriculture, A Global Perspective.* Proc. Int. Conf. on Dryland Farming, Amarillo/Bushland, TX, August 1988. Texas Agric. Exp. Stn., College Station.

Black, A.L. and J.K. Aase. 1988. The use of perennial herbaceous barriers for water conservation and the protection of soils and crops. *Agric. Ecosyst. Environ.* 22/23:135-148.

Black, A.L. and A. Bauer. 1983. Soil conservation: Northern Great Plains. In: H. E. Dregne and W. O. Willis (eds.) *Dryland Agriculture.* Agronomy 23:247-257.

Black, A.L. and A. Bauer. 1985. Soil conservation strategies for northern Great Plains. pp. 76-86. In: *.Planning and Management of Water Conservation Systems in the Great Plains States.* Proc. of a Workshop, Lincoln, Nebraska, October 1985. U. S. Dept. Agric., Soil Conserv. Serv., Lincoln, NE.

Black, A.L. and J.F. Power. 1965. Effect of chemical and mechanical fallow methods on moisture storage, wheat yields, and soil erodibility. *Soil Sci. Soc. Am. Proc.* 29:465-468.

Black, A.L. and F.H. Siddoway. 1976. Dryland cropping sequences within a tall wheatgrass barrier system. *J. Soil Water Conserv.* 31:101-105.

Black, A.L. and F.H. Siddoway. 1977. Winter wheat recropping on dryland as affected by stubble height and nitrogen fertilization. *Soil Sci. Soc. Am. J.* 41:1186-1190.

CTIC (Conservation Technology Information Center). 1990. Tillage definitions. *Conserv. Impact* 8(10):7.

Dickey, E., P. Jasa, and D. Shelton. 1990. Residue, tillage and erosion. pp. 73-83. In: *Conservation Tillage.* Proc. Great Plains Conserv. Tillage Symp., Bismarck, ND, August 1990. Great Plains Agric. Counc. Publ. 131.

Doran, J.W., W.W. Wilhelm, and J.F. Power. 1984. Crop residue removal and soil productivity with no-till corn, sorghum, and soybean. *Soil Sci. Soc. Am. J.* 48:640-645.

Finkel, H.J. 1986. Wind erosion. pp. 109-121. In: H. J. Finkel, M. Finkel, and Ze'ev Naveh (eds.) *Semiarid Soil and Water Conservation.* CRC Press, Inc., Boca Raton, FL.

Follett, R.F. and G.A. Peterson. 1988. Surface soil nutrient distribution as affected by wheat-fallow tillage systems. *Soil Sci. Soc. Am. J.* 52:141-147.

Fryrear, D.W. and E.L. Skidmore. 1985. Methods of controlling wind erosion. pp. 443-457. In: R.F. Follett and B.A. Stewart (eds.) *Soil Erosion and Crop Productivity.* ASA-CSSA-SSSA, Madison, WI.

Gerard, C.J. 1987. Methods to improve water infiltration on fragile soils. pp. 72-75. In: T.J. Gerik and B.L. Harris (eds.) *Conservation Tillage: Today and Tomorrow.* Proc. Southern Region No-Tillage Conf., College Station, TX, July 1987. Texas Agric. Exp. Stn. Misc. Publ. MP-1636, College Station.

Good, L.G. and D.E. Smika. 1978. Chemical fallow for soil and water conservation in the Great Plains. *J. Soil Water Conserv.* 33:89-90.

Greb, B. W. 1979. *Reducing drought effects on croplands in the west-central Great Plains.* U.S. Dept. Agric. Inf. Bull. 420. U.S. Gov. Print. Off., Washington, DC.

Greb, B.W., D.E. Smika, and A.L. Black. 1967. Effect of wheat straw mulch rates on soil water storage during summer fallow in the Great Plains. *Soil Sci. Soc. Am. Proc.* 31:556-559.

Haas, H.J., C.E. Evans, and E.F. Miles. 1957. *Nitrogen and carbon changes in Great Plains soils as influenced by cropping and soil treatments.* U. S. Dept. Agric. Tech. Bull. 1164. U.S. Gov. Print. Off., Washington, DC.

Haas, H.J., W.O. Willis, and J.J. Bond. 1974. Introduction. pp. 1-11. In: *Summer Fallow in the Western United States.* U. S. Dept. Agric., Agric. Res. Serv. Conserv. Res. Rpt. 17. U.S. Gov. Print. Off., Washington, DC.

Hagan, L.J. 1991. A wind erosion prediction system to meet user needs. *J. Soil Water Conserv.* 46:106-111.

Hallsted, A.L. and O.R. Mathews. 1936. *Soil moisture and winter wheat with suggestions for abandonment.* Kansas Agric. Exp. Stn. Bull. 273, Manhattan.

Havlin, J.L., D.E. Kissel, L.D. Maddux, M.M. Claassen, and J.H. Long. 1990. Crop rotation and tillage effects on soil organic carbon and nitrogen. *Soil Sci. Soc. Am. J.* 54:448-452.

Havlin, J.L. and A.J. Schlegel. 1990. Crop rotation and tillage effects on soil organic matter. pp. 225-232. In: *Conservation Tillage.* Proc. Great Plains Conserv. Tillage Symp., Bismarck, ND, August 1990. Great Plains Agric. Counc. Bull. 131.

Hobbs, J. A. and P. L. Brown. 1957. *Nitrogen and organic carbon changes in cultivated western Kansas soils.* Kansas Agric. Exp. Stn. Tech. Bull. 89, Manhattan.

Hobbs, J. A. and P. L. Brown. 1965. *Effects of cropping and management on nitrogen and organic carbon contents of a western Kansas soil.* Kansas Agric. Exp. Stn. Tech. Bull. 144, Manhattan.

Hooker, M. L., G. M. Herron, and P. Penas. 1982. Effects of residue burning, removal, and incorporation on irrigated cereal crop yields and soil chemical properties. *Soil Sci. Soc. Am. J.* 46:122-126.

Ivey, G. 1992. No-till cotton into wheat. Oral presentation, Great Plains Conserv. Tillage Workshop, Farwell, TX, March 1992.

Johnson, W. C., C. E. Van Doren, and E. Burnett. 1974. Summer fallow in the southern Great Plains. pp. 86-109. In: *Summer Fallow in the Western United States. U. S. Dept. Agric.*, Agric. Res. Serv. Conserv. Res. Rpt. 17. U. S. Gov. Print. Off., Washington, DC.

Jones, O. R. 1992. Water conservation practices in the Southern High Plains. pp. 22-25. In: Proc. Fourth Annual Conf. Colorado Conservation Tillage Assn., Sterling, CO, February 1992.

Jones, O.R., V.L. Hauser, and S.J. Smith. 1991. No-tillage effects on infiltration and runoff from a dry-farmed Torrertic Paleustoll. *Agron. Abstracts*, p. 333.

Laflen, J.M., L.J. Lane, and G.R. Foster. 1991a. WEPP, A new generation of erosion prediction technology. *J. Soil Water Conserv.* 46:34-38.

Laflen, J.M., W.J. Elliot, J. R. Simanton, C.S. Holzhey, and K.D. Kohl. 1991b. WEPP, Soil erodibility experiments for rangeland and cropland soils. *J. Soil Water Conserv.* 46:39-44.

Lamb, J.A., G.A. Peterson, and C.R. Fenster. 1985. Wheat fallow tillage systems' effect on a newly cultivated grassland soils' nitrogen budget. *Soil Sci. Soc. Am. J.* 49:352-356.

Lavake, D.E. and A.F. Wiese. 1979. Influence of weed growth and tillage interval during fallow on water storage, soil nitrates, and yield. *Soil Sci. Soc. Am. J.* 43:565-569.

Lyles, L., L.J. Hagan, and E.L. Skidmore. 1983. Soil conservation: Principles of erosion by wind. In: H.E. Dregne and W.O. Willis (eds.) *Dryland Agriculture*. Agronomy 23:177-188.

McCalla, T.M. and T.J. Army. 1961. Stubble mulch farming. *Adv. Agron.* 13:125-196.

Norwood, C.A., A.J. Schlegel, D.W. Morishita, and R.E. Gwin. 1990. Cropping system and tillage effects on available soil water and yield of grain sorghum and winter wheat. *J. Prod. Agric.* 3:356-362.

Phillips, W.M. 1964. A new technique of controlling weeds in sorghum in a wheat-sorghum-fallow rotation in the Great Plains. *Weeds* 12:42-44.

Phillips, W.M. 1969. Dryland sorghum production and weed control with minimum tillage. *Weed Sci.* 17:451-454.

Power, J.F., L.N. Mielke, J.W. Doran, and W.W. Wilhelm. 1984. Chemical, physical, and microbial changes in tilled soils. pp. 157-171. In: *Conservation Tillage*. Proc. Great Plains Conserv. Tillage Symp., North Platte, NE, August 1984. Great Plains Agric. Counc. Publ. 110.

Power, J.F., W.W. Wilhelm, and J.W. Doran. 1986. Crop residue effects on soil environment and dryland maize and soya bean production. *Soil Tillage Res.* 8:101-111.

Russel, J.C. 1939. The effect of surface cover on soil moisture losses by evaporation. *Soil Sci. Soc. Am. Proc.* 4:65-70.

Siddoway, F.H. and C.R. Fenster. 1983. Soil conservation: Western Great Plains. In: H.E. Dregne and W. O. Willis (eds.) *Dryland Agriculture*. Agronomy 23:231-246.

Skidmore, E. L., J.B. Layton, D.V. Armbrust, and M.L. Hooker. 1986. Soil physical properties as influenced by cropping and residue management. *Soil Sci. Soc. Am. J.* 50:415-419.

Smika, D.E. 1976. Seed zone soil water conditions with mechanical tillage in the semiarid Central Great Plains. pp. 1-6. In: Proc. Seventh Int. Soil Tillage Res. Org. Conf., Uppsala, Sweden. College of Agric. of Sweden, Uppsala.

Smika, D.E. and P.W. Unger. 1986. Effect of surface residues on soil water storage. *Adv. Soil Sci.* 5:111-138.

Smika, D.E. and G.A. Wicks. 1968. Soil water storage during fallow in the central Great Plains as influenced by tillage and herbicide treatments. *Soil Sci. Soc. Am. Proc.* 32:591-595.

Smith, S.J., A.N. Sharpley, J.W. Naney, W.A. Berg, and O.R. Jones. 1991. Water quality impacts associated with wheat culture in the Southern Plains. *J. Environ. Qual.* 20:244-249.

Stewart, B.A. 1990. Limits, bounds, and future of conservation tillage in the Great Plains. pp. 1-13. In: *Conservation Tillage.* Proc. Great Plains Conserv. Tillage Symp., Bismarck, ND, August 1990. Great Plains Agric. Counc. Bull. 131.

Tanaka, D.L. 1989. Spring wheat plant parameters as affected by fallow methods in the northern Great Plains. *Soil Sci. Soc. Am. J.* 53:1506-1511.

Unger, P. W. 1968. Soil organic matter and nitrogen changes during 24 years of dryland wheat tillage and cropping practices. *Soil Sci. Soc. Am. Proc.* 32:588-591.

Unger, P.W. 1972. Dryland winter wheat and grain sorghum cropping practices – Northern High Plains of Texas. Texas Agric. Exp. Stn. Misc. Publ. MP-933, College Station.

Unger, P.W. 1978. Straw-mulch rate effect on soil water storage and sorghum yield. *Soil Sci. Soc. Am. J.* 42:486-491.

Unger, P.W. 1982. Surface soil physical properties after 36 years of cropping to winter wheat. *Soil Sci. Soc. Am. J.* 46:796-801.

Unger, P.W. 1984a. Tillage and residue effects on wheat, sorghum, and sunflower grown in rotation. *Soil Sci. Soc. Am. J.* 48:885-891.

Unger, P.W. 1984b. Tillage effects on surface soil physical conditions and sorghum emergence. *Soil Sci. Soc. Am. J.* 48:1423-1432.

Unger, P.W. 1991. Organic matter, nutrient, and pH distribution in no- and conventional-tillage semiarid soils. *Agron. J.* 83:186-189.

Unger, P.W. 1992. Infiltration of simulated rainfall: Tillage system and crop residue effects. *Soil Sci. Soc. Am. J.* 56:283-289.

Unger, P.W., R.R. Allen, and A.F. Wiese. 1971. Tillage and herbicides for surface residue maintenance, weed control, and water conservation. *J. Soil Water Conserv.* 26:147-150.

Unger, P.W. and A.F. Wiese. 1979. Managing irrigated winter wheat residues for water storage and subsequent dryland grain sorghum production. *Soil Sci. Soc. Am. J.* 43:582-588.

USDA (U. S. Dept. Agric.). 1989. The second RCA appraisal. U. S. Gov. Print. Off., Washington, DC.

Webb, W.P. 1931. *The Great Plains.* Ginn and Company, Boston, New York, London, Atlanta, Dallas, Columbus, San Francisco.

Wicks, G. A. and D. E. Smika. 1973. Chemical fallow in a winter wheat-fallow rotation. *Weed Sci.* 21:97-102.

Wicks, G A., D.E. Smika, and G.W. Hergert. 1988. Long-term effects of no-tillage in a winter wheat (*Triticum aestivum*)-sorghum (*Sorghum bicolor*)-fallow rotation. *Weed Sci.* 36:384-393.

Wiese, A.F. and T.J. Army. 1958. Effect of tillage and chemical weed control practices on soil moisture storage and losses. *Agron. J.* 50:465-468.

Wiese, A.F. and T.J. Army. 1960. Effect of chemical fallow on soil moisture storage. *Agron. J.* 52:612-613.

Wiese, A.F., E. Burnett, and J.E. Box, Jr. 1967. Chemical fallow in dryland cropping sequences. *Agron. J.* 59:175-177.

Wilhelm, W.W., H. Bouzerzour, and J.F. Power. 1989. Soil disturbance-residue management effect on winter wheat growth and yield. *Agron. J.* 81:581-588.

Wilhelm, W.W., J.W. Doran, and J.F. Power. 1986. Corn and soybean yield response to crop residue management under no-tillage production systems. *Agron. J.* 78:184-189.

Willis, W.O., A. Bauer, and A.L. Black. 1983. Water conservation: Northern Great Plains. In: H.E. Dregne and W.O. Willis (eds.) *Dryland Agriculture.* Agronomy 23:73-88.

Willis, W.O. and A.B. Frank. 1975. Water conservation by snow management in North Dakota. pp. 155-162. In: Proc. Snow Management on the Great Plains Conf., Bismarck, ND, July 1975. Great Plains Agric. Counc. Publ. 73, Univ. Nebraska, Lincoln.

Wischmeier, W.H. and D.D. Smith. 1978. *Predicting rainfall erosion losses – A guide to conservation planning.* U. S. Dept. Agric. Handb. 537. U. S. Gov. Print. Off., Washington, DC.

Wood, C.W., D.G. Westfall, G.A. Peterson, and I.C. Burke. 1990. Impacts of cropping intensity on carbon and nitrogen mineralization under no-till dryland agroecosystems. *Agron. J.* 82:1115-1120.

Residue Management Strategies
for the Southeast

R.L. Blevins, W.W. Frye, M.G. Wagger, and D.D. Tyler

I. Introduction

Soil erosion has historically been a major problem confronting farmers in the Southeast. Conservationists long ago documented that large reductions in soil erosion can be achieved when all or part of the residue from previous crops is left on the soil surface. The need for appropriate residue management is especially critical for the Southeast because the soils are highly erodible and high energy rainstorms occur during the growing season.

A. Climate

The Southeast is characterized by a warm humid climate. The area has abundant energy from sunshine to promote vegetative growth. Most of the states have a precipitation distribution that often results in runoff, but still experience periods of soil water deficiencies in the summer when evapotranspiration exceeds

1-56670-003-5/94/$0.00+$.50

precipitation (Daniels et al., 1973). Soil temperature regimes range from the mesic zone in the northern sector to thermic farther south and hyperthermic in southern Florida. The climate is conducive to vegetative growth, with a growing season ranging from a minimum of 6 months to maximum of 12 months a year. Although the climate favors plant growth, it is difficult to maintain residues on the soil surface because of the warm moist climate that results in rapid decomposition of organic matter.

B. Soils

Highly weathered and well-developed Ultisols are the most extensive soils in the Southeast (Perkins et al., 1973). Soils in this order are acid, relatively infertile, and have low base saturation. These soils usually have low organic matter content and are easily dispersed. In many cases they are coarse-textured, particularly in the surface horizon. These conditions make the soils susceptible to erosion because the individual particles are easily detached, which is the first step in the soil erosion process. The Ultisols of the Atlantic Coastal Plain have been in continuous crops for the past 100 to 200 years. Because of degradation from erosion and poor management, many farms have been allowed to revert back to trees. This situation applies both to Coastal Plain and Piedmont provinces and, to some extent, certain other areas of the Southeast.

Along the western edge of the region, specifically the Mississippi Valley, both Alfisols and Ultisols are found. One common characteristic of this part of the Southeast is the presence of a layer of loess of varying thickness at the soil surface. In much of the area, the loess is 1 to several meters thick, but in other places, it is a thin layer overlying other parent material such as residuum of limestone, sandstone or shale. According to Langdale et al. (1985), these soils tend to be highly erodible due to their silty nature. Areas of western Kentucky and Tennessee have been identified as problem areas and targeted by the Soil Conservation Service as areas to concentrate erosion control efforts.

C. Crops of the Region

The major row crops of the Southeast include corn, wheat, and soybean. Cotton is a major crop in Alabama, Georgia, Mississippi, North Carolina and west Tennessee. In the Coastal Plain, peanut is an important crop, and vegetable and citrus production are important cash crops in Florida. Pasture and hay production in conjunction with livestock production involves a sizable portion of the land areas throughout the region. Although residue management is involved in forage production, the necessity of residue management isn't nearly as critical as under row crop production.

Table 1. Amount of corn, soybean, and cotton grown in the Southeast in 1991 using conservation tillage (>30% of surface covered with residue after planting)

State	Corn ha	%	Soybean Double crop ha	%	Soybean Single crop ha	%	Cotton ha	%
AL	26,071	22.7	32,817	63.4	25,330	14.9	5,075	3.0
FL	5,111	12.2	1,215	10.5	785	4.6	138	0.7
GA	43,956	16.9	41,301	36.0	42,675	23.1	7,526	4.3
KY	313,901	57.6	165,193	96.4	157,255	49.2	---	---
MS	24,478	32.6	41,623	54.4	173,840	23.8	34,495	7.2
NC	79,641	17.5	121,067	53.9	27,953	6.3	2,591	1.4
SC	12,870	11.8	21,765	21.1	6,137	12.4	1,395	1.7
TN	45,098	51.6	127,244	80.2	79,225	22.5	17,559	7.3
VA	141,057	61.1	96,846	94.8	56,663	43.9	---	---

(Adapted from Conservation Technology Information Center, 1991.)

II. Tillage Systems

Tillage systems in the Southeast range from the clean tillage provided by a moldboard plow to no-tillage where the soil and residue are left undisturbed except for a narrow band where the seeds are placed and covered. Intermediate conservation tillage methods (e.g. chisel plow) are used that leave at least 30% of the surface covered with residue after planting.

On the sandy coastal plain soils where traffic pans are common, deep plowing or subsoiling may be needed. A combination of in-row subsoiling and no-tillage planting has proven to be an effective approach to these particular soils.

For some cropping systems, there has been a rapid shift to reduced tillage (Table 1). For example, greater than 80% of the double-cropped soybeans following wheat harvest in Kentucky, Tennessee, and Virginia are grown no-tillage, and North Carolina, Mississippi, and Alabama are moving rapidly to this tillage/cropping system (CTIC, 1991). This system allows three crops to be harvested in two years, and provides adequate residue cover during the growing season when soil erosion potential is usually greatest. The northern most part of the region, which includes parts of the Appalachian region, had the highest percentage of both no-tillage and conservation tillage being used in 1991 (Table 1). This is encouraging since this part of the region requires conservation practices due to high erosion potential associated with the steeper slopes.

III. Managing Crop Residues

A. Removal

Residue management in the Southeast involves several approaches. One approach is removal by harvesting. This includes harvesting as hay or haylage or by grazing animals. Straw left after harvesting small grain crops such as wheat, barley, or oat is frequently baled and sold as bedding material. Another practice commonly used in the Southeast is to burn wheat straw and stubble before planting double-cropped soybean. From a sustainable agriculture viewpoint, these residues should be left on the soil surface.

B. Plow-Down or Chemical Burn-Down

Plowing under residues from the previous year's crop or existing green manure crop was the tillage system used almost exclusively in the Southeast before the 1970s. Since then, many farmers have shifted to conservation tillage methods. In Kentucky, Tennessee, and Virginia, most of the double-cropped soybean, for example, are no-tilled directly into small grain straw and stubble without any residue removal.

The plow-down method of managing residues mentioned above includes moldboard plowing with disking as secondary tillage, chisel-plowing plus disking, and disking only. Each system results in a different level of residue incorporation. In order to comply with the 1990 Farm Bill on highly erodible land, residues must be managed to leave a minimum of 30% of the surface covered after planting the row crop.

In most conservation tillage systems, herbicides are used to burn down existing vegetation, such as grass, weeds, or cover crops. Little or no incorporation is involved in the chemical burn-down/no-tillage system. In order to obtain adequate levels of residue to qualify as an effective erosion control practice, farmers in the Southeast often use winter annual cover crops to augment the sparse residue levels that carry over from crops such as peanut, soybean and cotton. In many cases, ordinary conservation tillage practices, such as chisel-plow and disk tillage, do not leave adequate cover where the previous year's stalks are the only residue remaining.

C. Residue Decomposition

The ability to maintain acceptable levels of residue at the soil surface is controlled by numerous factors. As mentioned above, climate is a key factor that determines both the intensity of cropping and organic matter decomposition rate. The hot, humid climate of the southern portion of the Southeast provides an environment where residue decomposes rapidly. Under these conditions, plant

Table 2. Decomposition of corn stalks as affected by tillage and placement

Year	Tillage	Placement	
		Surface	Buried
		------------------Weeks[a] ----------------	
1981	No-tillage	32	15
	Conventional	27	12
1982	No-tillage	27	11
	Conventional	22	16

[a]weeks required to reach the first half life of the residue.
(From Rice, 1983.)

materials with a high C:N ratio and lignin content, which in turn produce a longer lasting mulch, may be preferable. Corn residue and rye cover crops are examples of such plant materials, and both produce high levels of residue that decompose at slower rates than do legumes.

Rice (1983), in his studies on decomposition of corn stalks as affected by tillage and placement, concluded that placement of corn stalk residues had more influence on decomposition than tillage alone. Residues buried to a depth of 15 cm decomposed twice as fast as residues left on the surface (Table 2). Wilson and Hargrove (1986) showed that the rate of N disappearance from crimson clover residue was more rapid under conventional tillage than no-tillage conditions (Figure 1). They concluded that decomposition and N release were rapid enough to be of significant benefit to the summer crop.

A look at surface residue cover remaining approximately 3 weeks after corn planting under four tillage management methods in Kentucky is shown in Figure 2. This study site was seeded to a rye cover crop each year following corn harvest; therefore, the rye made a significant contribution to the mass of residue remaining as well as providing a better distribution of residue than corn stover alone. The shallow tandem disk tillage and deeper chisel-plow tillage with straight points (approximately 30 cm depth) resulted in 40 to 60% ground cover remaining, depending on nitrogen rates applied to the corn. No-tillage resulted in 95 to 98% cover compared to 2 to 5% cover where the moldboard plow was used. This is an example of a cropping system where high levels of crop residue are produced and residue material is somewhat slow to decompose.

IV. Cover Crops

An effective cropping strategy to assure adequate residue cover during the summer cropping season is to plant a cover crop in the fall. Annual cover crops of either legumes or nonlegumes are compatible with conservation tillage

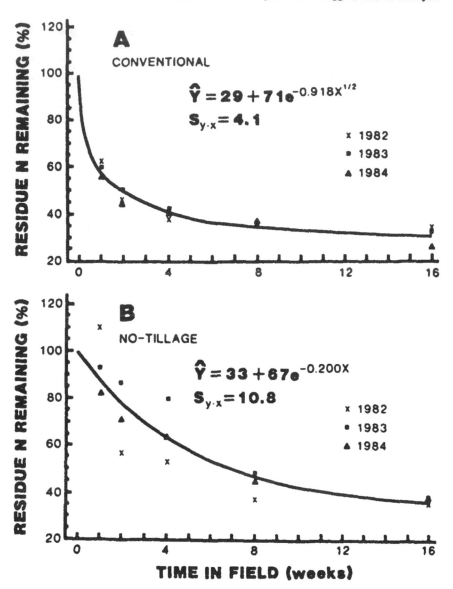

Figure 1. Percentage of original N remaining in crimson clover residue with time under conventional tillage (A) and no-tillage (B) conditions. (From Wilson and Hargrove, 1986.)

systems in the Southeast. Until recently, cover crops used in the Southeast have been mostly small grains, such as wheat and rye. Nonleguminous cover crops can provide sufficient vegetative cover to protect soils against water erosion during the off-season and the early part of the season of the principal row crop.

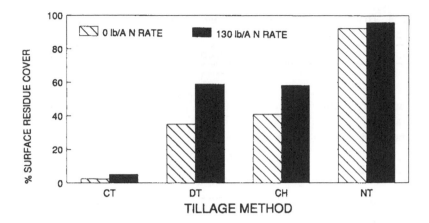

Figure 2. Percent surface residue measured three weeks after corn was planted. Residue includes previous corn residue and rye planted as a winter annual cover crop, Lexington, KY, 1989 (CT = conventional tillage, DT = disk tillage, CH = chisel-plow, NT = no-tillage).

Triplett (1986) reported that rye and wheat cover crops were especially effective, owing to the higher cellulose and lignin content and higher C:N ratios that make them resistant to decomposition.

Annual legume winter cover crops can provide soil and water conservation and supply a substantial part of nitrogen required for optimum yields of summer crops such as corn or grain sorghum (Ebelhar et al., 1984; Frye et al., 1988; Hargrove, 1986). Hairy vetch and crimson clover are legumes adaptable to such cropping strategies and are especially well adapted for the Southeast region. According to Bruce et al. (1991), cover crops have a great potential in restoration and maintenance of soil productivity, because they provide an on-site source of plant biomass that contributes to soil organic matter levels.

A. Managing for Corn and Soybean Production

Cover crops have been used for centuries to protect the soil against erosion, provide nutrients to the primary crop, and to enhance soil quality. In a classic field experiment in Virginia, Moschler et al. (1967) studied several winter crops in which corn was planted no-tillage. The results showed that rye was a highly efficient cover crop in producing dry matter (>7 Mg ha-1), and corn grain yields were significantly enhanced by the presence of any of the cover crops (Table 3). Gallaher (1977) also reported that corn yields increased in response to a rye cover crop.

In Kentucky, Munawar et al. (1990) reported positive effects on yields if rye cover crops were chemically desiccated or plowed down 2 to 3 weeks before

Table 3. Dry weight of cover crops and corn grain yield under no-tillage production in Blacksburg, VA

Cover crop	Cover crop (dry weight)	Corn grain yield
	----------------------Mg ha[-1]----------------------	
Rye	7.22	7.84a[1]
Rye and hairy vetch	8.55	7.17ab
Ryegrass	2.44	7.30ab
Ryegrass and crimson clover	2.82	6.72bc
Oat	2.06	5.82cd
Oat and vetch	2.73	5.40d
Weeds	2.39	5.26d

[1]Values followed by different letters are statistically different (P < 0.05).
(From Moschler et al., 1967.)

Table 4. Corn grain yield as affected by tillage and mulch treatments, averaged over all N rates, Lexington, KY, 1986

Tillage	Rye cover crop (early killed)	Rye cover crop (late killed)	Mean
	--------------------------Mg ha[-1]--------------------------		
Conventional	2.77	1,76	2.26c[1]
Disk	3.75	3.55	3.64b
Chisel-plow	4.29	3.77	4.03ab
No-tillage	4.61	4.20	4.41a
Mean	3.85a	3.32b	

[1]Mean values in a column or line followed by different letters are statistically different (P < 0.05).
(From Munawar et al., 1990.)

planting corn. Although the earlier killed treatment reduced rye dry matter by 40 to 50%, higher corn yields were observed for early killed rye (Table 4). They attributed the increased yields to reduced competition for soil water. When rye was allowed to continue growth to the grain-fill stage, substantial soil water was extracted as compared to early-kill of the rye.

Campbell et al. (1984a, 1984b) reported on an experiment in which corn and soybean were planted into a rye cover crop on a Norfolk loamy sand. During the 4-year study, the depletion of water by rye cover crops limited yields of the primary crops. They concluded that evapotranspiration by the rye had a more negative effect than the positive benefits the surface mulch provided in reducing evaporation during summer months. Soil water availability during the planting

Table 5. Cover crop dry matter production and N content as affected by time of burn-down in the North Carolina Piedmont

Cover crop and time of desiccation	Dry matter	N content
	Mg ha^{-1}	kg ha^{-1}
Rye		
Early	4.71	52
Late	7.33	66
Crimson clover		
Early	3.43	107
Late	4.98	131
Hairy vetch		
Early	2.35	114
Late	3.41	146

(From Wagger, 1989a.)

and early growth period of row crops being grown is the key to the use of and management of cover crops for no-tillage grain production.

Wagger's (1989a) work in North Carolina has provided valuable information on management strategies involving cover crops. Table 5 presents a comparison of dry matter production and N content of cover crops desiccated early versus late, two weeks apart. The late desiccation resulted in increased production of cover crop dry matter of 39% in rye, 41% in crimson clover, and 61% in hairy vetch. However, Wagger (1989b) concluded that cover crop management strategies should ensure that corn planting is not excessively delayed to allow for additional legume growth and N production.

Double-cropping soybean after wheat harvest is a popular and quite successful cropping system in several states in the Southeast where the growing season is long enough to accommodate the harvesting of two crops per year. Because of the large amounts of residue, double-cropping of grain crops is effective for controlling erosion (Mills et al., 1986). Research in Mississippi (Blaine et al., 1988) compared yield and net return for full season and double-cropped soybean relay planted into standing wheat with soybean no-till planted into wheat stubble (Table 6). The data show the importance of an earlier planting date for double-cropped soybean. If the wheat can be harvested early enough (mid-June), soybean can be planted in stubble without significant reductions in yield compared to full-season soybean. This is especially true if there are no severe water stress periods in late August or early September. In the Mississippi study, the best net return was realized by interplanting (relay planting) soybean into standing wheat about June 1. Small wheat yield losses occurred due to the interplanting operation, and soybean yields were high.

Date of soybean planting studies in Tennessee (Graves et al., 1978) showed that yields are reduced when soybean are planted after June 1. Graves et al.

Table 6. Two-year average grain yield and net return from wheat and soybeans at Northeast Branch of Mississippi Agricultural Experiment Station

Cropping system	Planting date for soybeans	Yield		Net return
		Wheat	Soybeans	
		------------Mg ha⁻¹-----------		$ ha⁻¹
Full season soybean	5/14	--	2.20	183
Double-Cropped				
Relay planted	5/14-5/15	2.55	2.22	195
	5/29-6/2	2.69	2.22	215
Stubble planted	6/17-6/21	2.82	2.15	183
	7/2-7/7	2.75	0.67	119

(From Blaine et al., 1988.)

(1980) found in a 3-yr study that soybean yields were reduced by 39% when no-till planted following wheat as compared to single crop soybean. Using the averages of two years in which there was adequate moisture, the no-till soybeans after wheat was 4.9% lower than single cropped soybean. Late (last of June) planted soybean in wheat stubble has a higher risk of failure during a dry summer than earlier planted (May) conventional till soybean.

B. Managing for Cotton Production

Cotton acreage is increasing in the Southeast. Historically, cotton production has generally caused severe soil erosion and overall soil degradation in the Southeast. Perhaps more than any other crop, cotton needs better residue management. Many cotton farmers in the Southeast are keenly interested in planting cotton no-tillage into winter annual cover crops.

Cover crops are needed in cotton production because the crop residue left from cotton is sparse and doesn't provide adequate protection of the soil surface. It is important that farmers adopt a system of cotton production that will allow them to grow continuous cotton and still comply with the conservation requirements of the 1990 Farm Bill.

Research in Tennessee indicates that no-tillage cotton is a viable and profitable production system. Tennessee researchers recommend no-till planting cotton into a wheat cover crop that received no nitrogen fertilizer and was killed two weeks earlier. They (John Bradley, personal communication) reported equal yields of cotton planted no-tillage as compared to conventional tillage (Table 7). Although no-tillage showed no yield advantage over conventional tillage, the

Table 7. Cotton lint yields from long-term no-tillage and conventional tillage plots at Agricultural Station at Milan in 1992.

	Tillage system	
	No-tillage	Conventional
	--------------------------kg ha^{-1}--------------------	
Wheat or rye	937	915
(11 year average)		
Cotton stubble	967	960
(9 year average)		

(John Bradley, personal communication.)

long-term advantage for reducing soil erosion on these highly erodible soils should strongly favor use of no-tillage.

Hoskinson et al. (1988) reported that cotton no-till planted into small grain residue required N fertilizer, but cotton no-tilled into hairy vetch required little or no N fertilizer. Similar results were observed in Alabama (Brown et al., 1985). They reported that when no-till cotton followed hairy vetch, yields were equal to the fall plow system in 2 of the 3 years, and N fertilizer requirements were reduced approximately 34 kg ha^{-1}.

V. Conclusions

The warm humid climate of the southeastern USA permits intensive cropping. Despite the high annual precipitation, droughts still limit crop yields. High-intensity rainstorms result in excessive erosion and runoff (Reicosky et al., 1977). Reasons for careful management of surface residues may vary from area to area within the region, but erosion control and conservation of soil moisture seem to be the concerns in most of the region. On the somewhat fragile, eroded Piedmont soils and steep soils of Appalachia, surface residue has to be an integral part of the cropping/tillage strategy.

The climate of the Southeast makes double-cropping a viable production system. When double-cropping includes two grain crops, the economic return is usually greater than when a grain crop and a forage crop are rotated. For this reason, wheat/soybean and other double-crop combinations have become popular in the Southeast. Because these systems produce large amounts of residue, they are effective in controlling erosion when residues are left on the soil surface.

Use of winter annual nonlegume or legume cover crops is compatible with conservation tillage systems for the purpose of maintaining surface residue cover to conserve water and reduce soil erosion losses. This may become a necessary cropping/residue management strategy for the Southeast.

Nonlegume cover crops, such as rye and wheat provide good surface cover during winter months and produce high levels of biomass that decomposes slowly due to their high C:N ratio. Problems include depletion of soil water before the primary crop is planted and immobilization of N. Fortunately, in many cases the N immobilization problem is overcome by the additions of optimum recommended rates of N fertilizer (Tyler et al., 1987). Depleting the soil water may produce problems with plant stands and seedling growth of summer crops. These cover crops, whether legumes or nonlegumes, must be managed to avoid excessive depletion of soil moisture prior to planting the primary crop. This may require killing the cover crop at least one week before planting the primary crop.

A legume cover crop provides biologically fixed N to the primary crop in addition to the benefits offered by nonlegume cover crops. The use of legume cover crops in the Southeast results in higher grain yields and shows promise for restoring soil productivity (Bruce et al., 1987).

Finally, because the soils and climate of the Southeast are conducive to excessive soil erosion, installing management strategies to effectively utilize residues from previous crops and cover crops should be our highest priority in the next few years.

References

Blaine, M.A., N.W. Buehring, J.G. Hamill, and D.B. Reginelli. 1988. Soybean-wheat intercropping response and effect on estimated net returns. *Proceedings 1988 Southern Conservation Tillage Conference,* Tupelo, MS. Special Bull. 88-1, August 1988. Mississippi State, MS 39762.

Brown, S.M., T. Whitwell, J.T. Touchton, and C.H. Burmester. 1985. Conservation tillage systems for cotton production. *Soil Sci. Soc. Am. J.* 49:1256-1260.

Bruce, R.R., S.R. Wilkinson, and G.W. Langdale. 1987. Legume effects on soil erosion and productivity. p. 81-89. In: J.F. Power (ed.) *The Role of Legumes in Conservation Tillage Systems.* Soil Conserv. Soc. Am., Ankeny, Iowa.

Bruce, R.R., P.F. Hendrix, and G.W. Langdale. 1991. Role of cover crops in recovery and maintenance of soil productivity. p. 109-114. In: W.L. Hargrove (ed.) *Cover Crops for Clean Water.* Soil and Water Cons. Soc., Ankeny, Iowa.

Campbell, R.B., D.L. Karlen, and R.E. Sojka. 1984a. Conservation tillage for maize production in the U.S. southeastern Coastal Plain. *Soil Tillage Res.* 4:511-529.

Campbell, R.B., D.L. Karlen, and R.E. Sojka. 1984b. Conservation tillage for soybeans in the U.S. southeastern Coastal Plain. *Soil Tillage Res.* 4:531-541.

Conservation Technology Information Center. 1991. *1991 National Survey of Conservation Tillage Practices*. West Lafayette, IN 47906-1334.

Daniels, R.B., B.L. Allen, H.H. Bailey, and F.H. Beinroth. 1973. Physiography. p. 3-16. In: S.W. Buol (ed.) *Soils of the Southern States and Puerto Rico*. Southern Coop. Series Bull. No. 174.

Ebelhar, S.A., W.W. Frye, and R.L. Blevins. 1984. Nitrogen from legume cover crops for no-tillage corn. *Agron. J.* 76:51-55.

Frye, W.W., J.J. Varco, R.L. Blevins, M.S. Smith, and S.J. Corak. 1988. Role of annual legume cover crops in efficient use of water and nitrogen. p. 129-154. In: W.L. Hargrove (ed.) *Cropping Strategies for Efficient Use of Water and Nitrogen*. ASA-CSSA-SSSA Special Publication No. 51. Madison, WI.

Gallaher, R.N. 1977. Soil moisture conservation and yield of crop no-till planted in rye. *Soil Sci. Soc. Am. J.* 41:145-147.

Graves, C.R., J.R. Overton, and H. Morgan. 1978. *Soybean--variety--Date of planting study from 1974-76*. Tennessee Agric. Exp. Stn. Bull 582, Knoxville, TN 37901.

Graves, C.R., T. McCutchen, L.S. Jeffrey, J.R. Overton, and R.M. Hayes. 1980. *Soybean-wheat cropping systems: Evaluation of planting methods, varieties, row spacings, and weed control*. Tennessee Agric. Exp. Stn. Bull. 597, Knoxville, TN 37901.

Hargrove, W.L. 1986. Winter legumes as a nitrogen source for no-till grain sorghum. *Agron. J.* 78:70-74.

Hoskinson, P.E., D.D. Tyler, and R.M. Hayes. 1988. Long term no-till cotton yields as affected by nitrogen. *Proceedings of Beltwide Cotton Production Research Conferences. National Cotton Council of America*. Memphis, TN.

Langdale, G.W., H.P. Denton, A.W. White, J.W. Gilliam and W.W. Frye. 1985. Effects of soil erosion on crop productivity of southern soils. p. 251-270. In: R.F. Follet and B.A. Stewart (eds.) *Soil Erosion and Crop Productivity*. American Society of Agronomy. Madison, WI.

Mills, W.C., A.W. Thomas, and G.W. Langdale. 1986. Estimating soil loss probabilities for Southern Piedmont cropping-tillage systems. *Trans. Am. Soc. Agr. Eng.* 29:948-955.

Moschler, W.W., G.M. Shear, D.L. Hallock, R.D. Sears, and G.D. Jones. 1967. Winter cover crops for sod-planted corn: Their selection and management. *Agron. J.* 59:547-551.

Munawar, A., R.L. Blevins, W.W. Frye, and M.R. Saul. 1990. Tillage and cover crop management for soil water conservation. *Agron. J.* 82: 773-777.

Perkins, H.F., H.J. Byrd, and F.T. Ritchie, Jr. 1973. Ultisols- light colored soils of the warm temperate forest lands. p. 73-86. In: S.W. Buol (ed.) *Soils of Southern States and Puerto Rico*. Southern Coop. Series Bull. No. 174.

Reicosky, D.C., D.K. Cassel, R.L. Blevins, W.R. Gill, and G.C. Naderman. 1977. Conservation tillage in the Southeast. *J. Soil Water Cons.* 32:13-19.

Rice, C.W. 1983. *Microbial nitrogen transformations in no-till soils*. Ph.D. diss. University of Kentucky, Lexington.

Triplett, G.B. 1986. Crop management practices for surface tillage systems. p. 149-182. In M.A. Sprague and G.B. Triplett (eds.). *No-tillage and surface tillage agriculture: The tillage revolution*. John Wiley & Sons, New York.

Tyler, D.D., B.N. Duck, J.G. Graveel, and J.F. Bowen. 1987. Estimating response curves of legume nitrogen contributions to no-till corn. In: J.F. Power (ed.). *The role of legumes in conservation tillage systems*. Proceedings of a National Conference. Athens, GA. April 27-29, 1987. Soil Conservation Society of America, Ankeny, Iowa.

Wagger, M.G. 1989a. Time of desiccation effects on plant composition and subsequent nitrogen release from several winter annual cover crops. *Agron. J.* 81:236-241.

Wagger, M.G. 1989b. Cover crops management and nitrogen rate in relation to growth and yield of no-till corn. *Agron. J.* 81:533-538.

Wilson, D.O. and W.L. Hargrove. 1986. Release of nitrogen from crimson clover residue under two tillage systems. *Soil Sci. Soc. Am. J.* 50:1251-1254.

Residue Management Strategies in the Northeast

J.K. Radke and C.W. Honeycutt

I. Introduction

The Northeast is unique in many ways when compared to the rest of the United States. Agriculture in this area has its own flavor and problems associated with crop residues and farm management systems. Residue management on agricultural land in the Northeast often reflects the preponderance of population

MID ATLANTIC STATES   NEW ENGLAND STATES

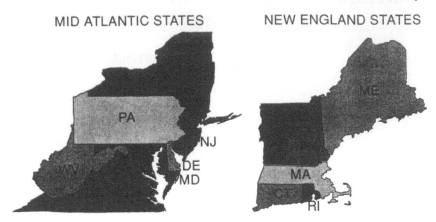

Figure 1. Map of the Mid-Atlantic and New England states of the Northeast.

density and industrialization. Cities and industries are looking more towards agricultural land for waste disposal because other disposal means are becoming limited, highly regulated, and costly.

II. Characteristics of the 13 Northeast States and the District of Columbia

The northeastern states range from Maine in the north to Virginia in the south and include the entire urban District of Columbia. Figure 1 shows the Northeast divided into the New England and Mid-Atlantic regions. Maine is agriculturally unique among the New England states and West Virginia differs agriculturally from the other Mid-Atlantic states. Several characteristics of the Northeast are pertinent to residue management. These include population density, geology, soils, climate, topography and agricultural impact on the economy as detailed below.

A. Population

In 1990, 25% (64.2 million) of the total U.S. population (260 million) lived in the Northeast. Population density ranged from 3200 people/km^2 in the District of Columbia to a relatively sparse 14 people/km^2 in Maine (Table 1). Many northeast states are highly urbanized. In Table 1, the population is defined as urban if less than 50% of the family income is derived from farming, forestry, or mining. This gives rise to the 100% urbanization in New Jersey although

Table 1. Area and population statistics of the Northeast region, 1990

State	Area	Population	Population density	Urban population
	(1000 ha)	(1000's)	(number km^{-2})	(%)
CT	1,300	3,230	248.3	92.6
DE	530	660	124.2	65.9
DC	20	580	3,217.5	100.0
ME	8,620	1,220	14.1	36.1
MD	2,710	4,730	174.7	92.9
MA	2,150	5,930	276.3	90.6
NH	2,400	1,100	45.9	56.3
NJ	2,020	7,810	387.1	100.0
NY	12,720	17,630	138.6	91.2
PA	11,730	11,870	101.1	84.8
RI	310	990	315.1	92.6
VT	2490	560	22.5	23.2
VA	10,560	6,130	58.0	72.2
WV	6,280	1,780	28.4	36.5
Total	63,840	64,220		
Mean			368.0	73.9

(Data from PC GLOBE, Inc., 1990.)

substantial farming exists in that state. Many hobby farms, truck farms, and large gardens are found in New Jersey.

B. Geology

The geologic history of the Northeast dates back to the Precambrian Grenville orogeny, a period ending approximately 950 million years ago (Ciolkosz et al., 1984). The geologic record since that time provides evidence of repeated epochs of tectonism, sedimentation, folding, faulting, warping, metamorphism, intrusion, volcanism, and glaciation. The most recent geologic event in the region was the Pleistocene glaciation (Flint, 1971). During this time ice covered approximately half of the region (Figure 2) (Ciolkosz et al., 1984). Although periglacial conditions may have existed south of the glacial border, this border represents a significant demarcation in soil parent material and time of weathering for soils of the Northeast.

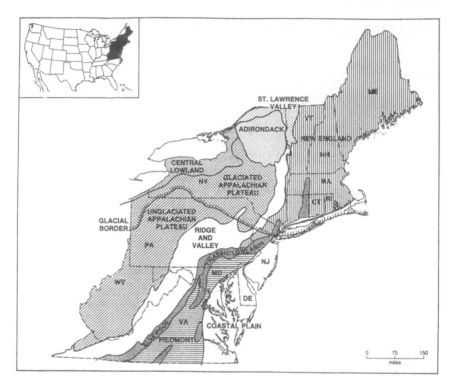

Figure 2. Physiographic map of the Northeast. (From Ciolkosz et al., 1984.)

C. Soils

Eight of the 11 soil orders currently recognized in the U.S. soil taxonomic
system occur in the northeastern U.S. (Smith, 1984). Most soils in this region
(approximately 38%) are Inceptisols with 29% belonging to the Ochrept
suborder and the remainder found as Aquepts (Miller and Quandt, 1984).
Inceptisols exhibit limited horizon alteration. Their minimal pedogenic development may result from geologically young sediments or landscapes (e.g. erosion
induced) or from factors such as resistant parent material, low temperature, or
high water table (Foss et al., 1983).

Ochrepts are primarily light-colored, brownish, and freely drained (Soil
Survey Staff, 1975). Ochrepts south of northern Pennsylvania are primarily low
in base saturation (Dystrochrepts) (Soil Survey Staff, 1975). Ochrepts in
southern New York and northern Pennsylvania commonly possess a fragipan
which restricts root development and water percolation. These soils are
primarily forested but are also used for hay, pasture, and some grain production
(Soil Survey Staff, 1975).

Aquepts are characterized by poor natural drainage and are commonly found with mottled gray and rusty colors. These soils are often found in depressions and floodplains. Ochrepts in New York and New Hampshire have been reported on upper landscape positions and to grade into Aquepts at the lower footslope and toeslope positions (Hanna et al., 1975; Gile, 1958).

Ultisols occupy approximately 24% of the northeast region. These are highly weathered soils exhibiting a subsoil accumulation of translocated clay and a low base saturation. Most Ultisols occur south of the glacial border (Figure 2). Ultisols are used for growing a wide variety of crops including corn, soybeans, tobacco, feed grains, hay, and forests (Soil Survey Staff, 1975).

Approximately 21% of northeastern soils are classified as Udults (Miller and Quandt, 1984). These are freely drained Ultisols with only a short or no marked dry season and a low organic matter content (Soil Survey Staff, 1975). Udults are found extensively on the Coastal Plain of Virginia, Maryland, Delaware, and New Jersey; the Piedmont section of Virginia and Maryland; and in the Unglaciated Appalachian Plateau of West Virginia and Pennsylvania (Figure 2).

Aquults are found in only about 3% of the region. These are poorly drained Ultisols that primarily occur in the Chesapeake Bay area and to a lesser extent in the Delaware Bay area (Smith, 1984).

Spodosols occupy approximately 16% of the Northeast (Miller and Quandt, 1984). These soils usually possess a horizon marked by accumulation of organic matter and aluminum. Most Spodosols possess little silicate clay. Their cation exchange capacity is primarily related to organic matter content and is therefore pH dependent (Yeck et al., 1984). Spodosols are naturally infertile soils but can be highly responsive to proper management (Soil Survey Staff, 1975).

Approximately 93% of northeastern Spodosols are Orthods. These are the relatively well-drained Spodosols characterized by iron, aluminum, and organic matter accumulation. Orthods occur extensively in Maine, New Hampshire, and Vermont (Smith, 1984). Aquods, the wet Spodosols, and Orthods occupy most of the Adirondack physiographic province (Figure 2).

Alfisols are found in approximately 11% of the northeast region. These are generally fertile soils characterized by the presence of an argillic horizon (illuvial clay accumulation), moderate to high base saturation, and the presence of plant-available water for at least three months during the growing season (Soil Survey Staff, 1975).

Approximately 75% of the northeast Alfisols are Udalfs (Miller and Quandt, 1984). These are relatively well drained soils, occupying extensive areas in the western Unglaciated Appalachian Plateau of West Virginia and Pennsylvania, the Central Lowland in New York, and the band of Triassic Lowlands of Virginia, Maryland, Pennsylvania, and New Jersey (Figure 2) (Smith, 1984). The poorly drained Alfisols (Aqualfs) are found in the St. Lawrence Valley, the Glaciated Appalachian Plateau of western Pennsylvania, and in scattered associations with other soils in the Northeast (Smith, 1984).

Entisols, Histosols, Mollisols, and even Vertisols are also found in the Northeast. These soils occupy approximately 5, < 2, < 1, and < 1% of the region, respectively (Miller and Quandt, 1984; Smith, 1984).

D. Climate

Climatic parameters vary greatly within the Northeast. Mean length of the frost-free period ranges from 90 days in the northern mountains to 240 days at the mouth of the Chesapeake Bay (USDC, 1968). Soil temperature regimes, generally measured at 50 cm for classification purposes, range from cryic (mean annual temperature = 0 to 8°C) in northern Maine and some mountainous regions to thermic (mean annual temperature = 15 to 22°C) in eastern Maryland and Virginia (Smith, 1984; Soil Survey Staff, 1975). Mean annual precipitation varies considerably, with values ranging from approximately 800 to 1320 mm/yr (USDC, 1968).

E. Topography and Tilled Land

Much of the terrain in the Northeast is rugged to mountainous and not suitable for row-crop agriculture. Therefore, farming and other agricultural pursuits are often relegated to flatter river valleys. Considerable areas of land are covered with forests or are used for non-agricultural purposes. Only Delaware and Maryland have over 20% of the land cropped while four states have less than 1% cropped (Table 2).

F. Agricultural Impact

Agriculture plays an important role in the economy of the Northeast, but is not as prominent as in other regions of the U.S. Pennsylvania, New York, Virginia, and Maryland generated over one billion dollars from farm sales in 1989 (Table 3). However, only Pennsylvania (19th) and New York (23rd) were ranked in the top half of the nation for gross farm income in 1989, and nine northeast states ranked 39th or lower. Most northeast states received less than 2% of their gross income from farming and agricultural products.

It is important to note the independence of farmers in the Northeast. Only 2% of the total government payments went to the Northeast in 1989. Only one of these states received over 10% of its 1989 net farm income from government payments in contrast to some midwest states where the average farm would have operated at a loss without the government payments (USDA-ERS, 1991).

The Northeast compared to the nation has 25% of the population, 7% of the land area, 3.9% of the cropland, and 8% of the gross farm income. Incomes from crops and livestock are 6 and 10%, respectively, of the national total.

Table 2. Total area, cropped area and percentage of each northeast state cropped in 1991

State	Total area	Cropped area	
	(ha)	(ha)	(%)
CT	1,300,000	14,670	1.1
DE	530,000	219,550	41.4
DC	20,000	0	0.0
ME	8,620,000	84,290	1.0
MD	2,710,000	623,750	23.0
MA	2,150,000	30,720	1.4
NH	2,400,000	9,900	0.4
NJ	2,020,000	182,270	9.0
NY	12,720,000	894,010	7.0
PA	11,730,000	1,219,210	10.4
RI	310,000	2,580	0.8
VT	2,490,000	49,450	2.0
VI	10,560,000	782,620	7.4
WV	6,280,000	58,810	0.9
Northeast	63,840,000	4,171,830	6.5

(Data from PC GLOBE, Inc., 1990 and Conservation Technology Information Center, 1991.)

Income from livestock is greater than from crops in the Northeast. This has important implications for residue management. Animal manures need to be factored into farm management systems because they not only provide valuable nutrients and organic matter, but could also become pollutants if not used wisely.

All northeast states list dairy as one of their top five commodities (Table 4). Cattle and poultry also rank high on the list for many of the states. The greenhouse and nursery industries are very important in several states. Some states are national leaders in the production of specialty crops such as cranberries, maple products, mushrooms, and wild blueberries.

III. Residue Materials

Residue materials used for mulching or soil amendments in the Northeast come from off-farm as well as on-farm sources. On-farm sources include crop residues left in the field; cover and green manure crops grown for soil conservation, organic matter, and nitrogen; and animal manure with or without bedding materials. Off-farm materials include processing and urban wastes. Processing wastes largely originate from the papermill and lumber industries or the food and feed processing industries. These materials include papermill

J.K. Radke and C.W. Honeycutt

Table 3. Northeast farm income from gross receipts, marketing receipts, crop sales, livestock sales and government payments, and the national rankings for each category in 1989

State	Gross	RK¹	Marketing	RK¹	Crop	RK¹	Livestock	RK¹	Govt.	RK¹
	million $		million $		million $		million $		million $	
CT	435	45	404	45	218	41	186	44	2	47
DE	699	40	663	39	160	44	503	39	5	44
ME	522	42	447	42	233	40	215	42	7	41
MD	1,502	35	1,346	35	476	36	870	31	24	38
MA	480	44	429	43	317	39	112	46	4	45
NH	158	48	142	48	79	46	63	48	2	46
NJ	767	39	660	40	463	37	197	43	22	39
NY	3,284	23	2,857	23	911	28	1,946	17	76	29
PA	4,009	19	3,581	19	986	26	2,595	10	68	31
RI	90	49	79	49	66	47	13	49	0	50
VT	481	43	426	44	51	49	375	40	7	42
VA	2,487	29	2,058	29	685	33	1,372	22	39	34
WV	414	46	314	46	64	48	250	41	12	40
Total	15,328		13,406		4,709		8,697		268	
U.S.	189,219		159,173		75,449		83,724		10,887	

¹ RK = National rank. (Data from U.S. Department of Agriculture, ERS, 1991.)

Table 4. The top five commodities produced by each northeast state in 1990[1]

State	Ranking				
	1	2	3	4	5
CT	Grnh/Nurs	Eggs	Dairy Prod	Tobacco	Cattle/Calf
DE	Broilers	Soybeans	Grhn/Nurs	Corn	Dairy Prod
ME	Potatoes	Eggs	Dairy Prod	Apples	Grhn/Nurs
MD	Broilers	Dairy Prod	Grhn/Nurs	Soybeans	Cattle/Calf
MA	Grhn/Nurs	Cranberry	Dairy Prod	Eggs	Apples
NH	Dairy Prod	Grhn/Nurs	Apples	Cattle/Calf	Hay
NJ	Grhn/Nurs	Dairy Prod	Eggs	Blueberry	Peaches
NY	Dairy Prod	Grhn/Nurs	Cattle/Calf	Apples	Hay
PA	Dairy Prod	Cattle/Calf	Grhn/Nurs	Eggs	Mushroom
RI	Grhn/Nurs	Dairy Prod	Eggs	Potatoes	Apples
VT	Dairy Prod	Cattle/Calf	Hay	Maple Prod	Grhn/Nurs
VA	Cattle/Calf	Broilers	Dairy Prod	Tobacco	Turkeys
WV	Cattle/Calf	Broilers	Dairy Prod	Turkeys	Apples

[1]Grhn = Greenhouse; Nurs = Nursery; Prod = Products.
(Data from U.S. Department of Agriculture, ERS, 1991.)

sludge, wood chips, wood ash, potato processing residues, fish processing remains, and many other organic residues suitable for disposal on agricultural land. Urban sources include yard waste (grass clippings, leaves, and tree trimmings), sewage sludge, and organic trash.

A. Crop Residues

A production summary for several major crops grown in the Northeast is shown in Table 5. It is difficult to accurately determine the amount of crop residues left on the land each year. From data in the 1987 Census of Agriculture (USDC, 1989), the amount of residue produced (Table 6) was estimated based on the amount of grain harvested for corn, wheat, oats, and barley and the area harvested for corn silage. Much of the straw from wheat and oats is harvested for animal bedding or for sale off the farm. Corn grown for silage leaves little residue remaining in the field. For perspective, Inglett (1973, p.124) reported that the Northeast produced about 3.9 million mt yr^{-1} of cereal straw (about 3% of the U.S. total). Pennsylvania alone produced over 41% of the Northeast's total cereal crop residue and together with New York, Virginia, and Maryland produced over 90% of the Northeast's total in the early 1970s.

Only Delaware, Virginia, and Maryland had over 50% of their cropland in conservation tillage (greater than 30% residue left on the surface) in 1991 (NACD-CTIC, 1991). Six of the thirteen northeast states had over one-third of their cropland in conservation tillage. Cover and green manure crops are grown

Table 5. Production of corn in 1987 and soybeans, wheat, oats, potatoes, and grapes in 1988 for the Northeast

State	Corn	Soybeans	Wheat	Oats	Potatoes	Grapes
			mt			
CT	8,460	0	0	0	6,140	0
DE	224,010	165,680	89,350	0	82,090	0
ME	0	0	0	45,210	1,000,000	0
MD	800,220	384,680	245,730	13,660	20,910	0
MA	0	0	0	0	26,000	0
NH	0	0	0	0	0	0
NJ	190,100	75,850	38,050	3,090	46,090	0
NY	1,674,940	0	135,000	116,530	308,730	142,730
PA	2,526,090	196,360	245,730	200,910	167,730	52,730
RI	190	0	0	0	13,640	0
VT	25,770	0	0	0	910	0
VA	521,630	423,820	283,640	9,830	93,090	0
WV	82,750	0	11,290	4,450	0	0
Total	6,054,160	1,246,390	1,048,790	393,680	1,765,330	195,460

(Corn data from U.S. Department of Commerce, 1989; all other data from PC GLOBE, Inc., 1990.)

in many areas of the Northeast and are important components in farming systems using crop rotations, but general statistics are not available.

B. Animal Manure

Certain areas in the Northeast produce large amounts of animal manure. The amount of manure produced can exceed the local area available for landspreading in reasonable quantities. However, there are many soils that can greatly benefit from manure addition.

No data was found on the amount of manure produced by farm animals in the Northeast. Consequently, we took the number of cattle, hogs, sheep, and chickens (Table 7) and estimated manure production using factors given by Morrison (1957). Manure production by dairy cows was estimated from milk production using a similar procedure. Estimated manure production for each state is presented in Table 8. Manure is a significant "residue" in several northeast states.

Table 6. Estimated residue produced by crops grown in the Northeast[1,2]

State	Grain corn	Silage corn	Wheat	Oats	Barley
			mt		
CT	8,550	0	100	150	0
DE	251,340	4,040	69,810	870	44,710
ME	0	10,440	980	67,960	950
MD	812,990	33,420	265,770	19,910	86,450
MA	0	10,350	240	200	0
NH	0	0	0	150	0
NJ	192,710	8,330	34,150	0	12,660
NY	1,677,510	191,080	0	238,470	0
PA	2,527,730	159,350	301,130	347,260	80,920
RI	190	720	0	0	0
VT	25,770	25,550	590	0	1,300
VA	532,730	79,150	316,680	12,790	0
WV	82,880	13,210	12,420	5,690	0
Total	6,112,400	535,640	1,001,870	693,450	226,990

[1] Estimated from area and production data in the 1987 Census of Agriculture (USDC, 1989).
[2] States sometimes reported growing small areas of specific crops, yet failed to provide yield data. Crop residue was considered to equal zero in these cases.

C. Processing Waste

Some processing wastes are suitable for disposal on agricultural land. Some of the major concerns are the presence of heavy metals, toxic chemicals, pests, or disease-causing organisms. Wastes with high contents of needed nutrients and/or organic matter can be used to benefit crops. Processing wastes are varied and often localized. For example, fish processing is usually conducted on the coast. Potato processing is conducted in Maine, Massachusetts, and Connecticut. Papermill wastes are generated where major papermills are located. However, the potential for waste generation to exceed agricultural land disposal capacity is a definite possibility, especially in states with a large number of these industries and limited agricultural land. Seekins and Mattei (1990) presented 1989 data (Table 9) showing how various waste materials were utilized or discarded in Maine. In 1990, various municipal and processing wastes were landspread on 4400 hectares in Maine (Table 10). Industrial ash required over 50% of the land used for disposal in this study.

Table 7. Livestock and milk produced in the Northeast, 1988

State	Cattle	Hogs	Sheep	Chicken	Milk
	(1000 head)				(1000 mt)
CT	70	10	10	4,190	250
DE	30	30	0	610	60
DC	0	0	0	0	0
ME	110	10	20	4,710	290
MD	330	170	40	3,540	660
MA	70	30	10	960	210
NH	60	10	10	210	140
NJ	80	40	20	1,700	180
NY	1,580	150	80	4,150	5,190
PA	1,920	970	130	20,000	4,640
RI	10	10	0	180	20
VT	300	10	20	160	1,090
VA	1,670	400	120	3,670	910
WV	500	40	80	470	140
Total	6,730	1,880	540	44,550	13,780

(Data from PC GLOBE, Inc., 1990.)

Table 8. Estimated manure produced by farm animals in the Northeast, 1988

State	Cattle	Dairy	Hogs	Sheep	Chicken
	(1000 mt)				
CT	518	721	11	5.6	54.4
DE	206	180	53	0.0	7.9
ME	746	858	13	10.5	61.2
MD	2,329	1,924	272	24.5	46.0
MA	497	617	50	9.8	12.5
NH	391	404	14	7.7	2.7
NJ	533	528	61	10.5	22.1
NY	11,246	15,231	242	56.7	54.0
PA	13,632	13,602	1,552	93.8	260.0
RI	43	52	10	0	2.4
VT	2,109	3,182	10	11.9	2.1
VA	11,857	2,666	640	80.5	47.7
WV	3,550	401	59	57.4	6.1
Total	47,657	40,366	2,987	368.9	579.1

(Calculated from data in PC GLOBE, Inc., 1990.)

Table 9. Summary of waste production and utilization in Maine, 1989

Type of Waste	Landfilled		Landspread		Composted		Other		Total
	Quantity	%	Quantity	%	Quantity	%	Quantity	%	Quantity
Papermill sludge (mt)	663,200	65	23,800	2	-	-	341,200	33	1,028,200
Municipal sludge:									
Solid (mt)	52,800	41	34,400	27	17,600	14	22,600	18	127,400
Liquid (m³)	2,290	4	49,600	88	2,150	4	2,400	4	56,400
Seafood waste (mt)	-	-	-	-	3,100	16	16,000	84	19,100
Animal manure:									
Stables (mt)	2,360	25	6,010	63	-	-	1,140	12	9,510
Fairs (mt)	1,120	13	2,860	34	3,850	46	560	7	8,390
Tracks (mt)	0	0	4,310	61	2,790	39	0	0	7,100
Wood ash (mt)	75,800	55	61,300	44	-	-	1,350	1	138,450
Vegetable wastes:									
Solid (mt)	435	1	28,500	32	2,450	3	56,700	64	88,085
Liquid (m³)	0	0	22,220	77	-	-	6,620	23	28,840

(Summarized from data in Seekins and Mattei, 1990.)

Table 10. Maine waste utilization, 1990[1,2]

Waste item	Amount used	Broadcast area
	Amount ha[-1]	ha
Municipal Sludge		
- Liquid	140,310 l	357
- Solid	28.3 m³	902
Septage	374,160 l	89
Ash	37.2 mt	2,440
Pulp and Papermill Sludge	403.5 mt	310
Fish		
- Liquid	93,540 l	7
- Scales	4.4 mt	59
Potato		
- Liquid	351,900 l	121
- Solid	94.5 m³	116

[1] Personal Communication: Maine Department of Environmental Protection Memo on Sludge and Residual Utilization.
[2] Pulp and papermill sludge were spread on forest land. Remaining wastes are thought to have been spread on agricultural land.

D. Urban Waste

As landfill space becomes scarce, ocean dumping is eliminated, and incineration capacity is exceeded or becomes too costly, municipalities are looking more closely at disposal on agricultural lands. Paper trash from a typical town in the Northeast makes up a large percentage of the municipal waste (Figure 3). Yard wastes (grass clippings, leaves, and tree trimmings) constitute a large volume of municipal waste and are of particular concern because of their bulk. Such bulky wastes often need to be processed before being spread on farm fields. Sewage sludge and effluent are also candidates for land disposal provided certain precautions are taken.

IV. Residue Handling

Perhaps the key to residue handling is to recycle whenever possible. Crop residues and the nutrients they contain are recycled if left in the field or later returned to the field as animal manure, compost, or waste products. More

Municipal Waste

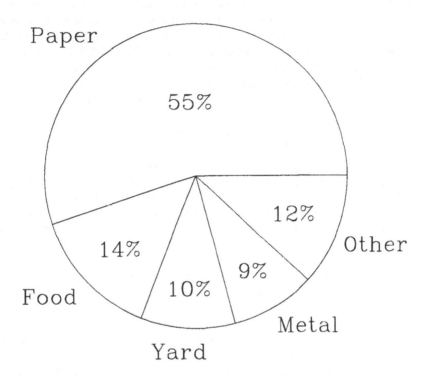

Figure 3. Breakdown of municipal waste from a typical town in the Northeast.

municipal waste needs to be recycled rather than dumped in landfills or incinerated. As mentioned earlier, ocean dumping should not be tolerated. Possible uses of waste products include animal feed, animal bedding, biogas production, heat production, composting material, fertilizer, soil amendments, and other saleable items. Considerable handling, processing, and transportation may be required. Many states and municipalities are developing stringent rules and regulations regarding the handling, processing, and disposal of wastes to protect people and the environment. As disposal costs continue to rise, alternative means of safe disposal are pursued, which often means increased disposal on agricultural lands.

V. Conservation Tillage in the Northeast

Conservation tillage (CT) is defined as "any tillage and planting system that maintains at least 30% of the soil surface covered by residue after planting to reduce soil erosion by water; or where soil erosion by wind is the primary concern, maintains at least 1000 pounds of flat, small grain residue equivalent on the surface during the critical erosion period" (NACD-CTIC, 1991). Data presented in the 1991 National Survey of Conservation Tillage Practices is categorized into no-till (NT), ridge-till (RT), mulch-till (MT), 15 to 30% residue cover, and less than 15% residue cover (NACD-CTIC, 1991). Data for the northeast states are summarized in Table 11. Over half of the total tilled land in the Northeast has less than 15% cover. About 20% of cultivated land is in NT and 17% in MT. Less than 0.1% is in RT. The New England states with the exception of Maine have very little cultivated land under CT (Figure 4a).

Despite these low percentages, CT adoption continues to increase in the Northeast with 37% of the tilled land in CT during 1991. Delaware had almost two-thirds of its tilled land in CT in 1991 (Figure 4b). Other states with more than one-third of tilled land in 1991 CT were, in decreasing order: Virginia, Maryland, Maine, West Virginia, and Pennsylvania. Maine was the only New England state in this group.

As discussed in more detail later, residue often increases soil water content and decreases soil temperature which can be detrimental in the northern states. Research in New York has shown that RT may provide an economic advantage over other CT methods (Cox et al., 1992). They found RT to be economical on all types of soils tested while NT was only economical on the well-drained soils. Cox et al. (1990) earlier reported that RT gave higher yields than NT in New York due to warmer soil temperatures. Increased use of RT in New York is desirable since less than 3% of the agricultural land is no-tilled and only 15% is mulch-tilled. Ridge-till has a disadvantage in New York because it is more difficult to use with sod crops (Cox et al., 1992). Other groups including the Rodale Research Center near Kutztown, Pennsylvania (S.E. Peters, 1993 personal communication) are experimenting with small grain, hay crops and cover crops in RT systems. As various problems with RT are solved, more land will likely be ridge-tilled in the Northeast.

VI. Residue Management Considerations for Nutrient Cycling

Tillage is an important tool for crop residue, manure, and waste management. It is clear from previous sections that the northeast region covers a wide range of soils; varying in texture, drainage, mineralogy, and effective rooting depth; and a considerable range in climate. Thus, the impact of tillage systems on crop residue decomposition and nutrient cycling is perhaps best examined by

Table 11. Land area under various tillage schemes in the Northeast, 1991

State	No-Till		Ridge-Till		Mulch-Till		Residue				Total Cropped
							15-30 %		< 15 %		
	ha	%	ha	%	ha	%	ha	%	ha	%	ha
CT	1,580	10.8	0	0.000	200	1.4	1,420	9.7	11,460	78.1	14,670
DE	79,860	36.4	0	0.000	65,450	29.8	8,870	4.0	65,370	29.8	219,550
ME	170	0.2	36	0.043	33,650	39.9	16,770	19.9	33,670	39.9	84,290
MD	204,830	32.8	231	0.037	113,550	18.2	95,250	15.3	209,890	33.6	623,750
MA	700	2.3	0	0.000	1,220	4.0	5,720	18.6	23,090	75.2	30,720
NH	250	2.5	0	0.000	580	5.9	290	2.9	8,780	88.7	9,900
NJ	29,970	16.4	81	0.044	10,100	5.5	21,140	11.6	120,970	66.4	182,270
NY	25,680	2.9	277	0.031	130,100	14.6	73,490	8.2	664,470	74.3	894,010
PA	221,220	18.1	417	0.034	206,600	16.9	195,700	16.1	595,280	48.8	1,219,210
RI	160	6.2	0	0.000	50	2.1	930	35.9	1,440	55.9	2,580
VT	3,010	6.1	0	0.000	940	1.9	920	1.9	44,580	90.1	49,450
VA	241,640	30.9	0	0.000	158,740	20.3	83,830	10.7	298,420	38.1	782,620
WV	17,430	29.6	12	0.020	4,880	8.3	7,040	12.0	29,440	50.1	58,810
Total	826,500	19.8	1,054	0.025	726,060	17.4	511,370	12.3	2,106,860	50.5	4,171,830

(Data from National Association of Conservation Districts, CTIC, 1991.)

New England Tilled Land (1991)
Total: 191,611 Hectares

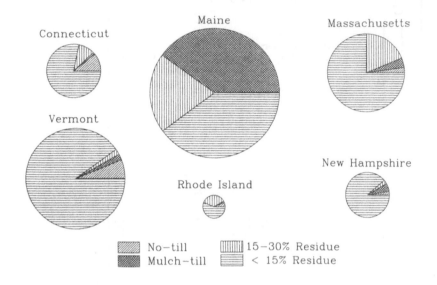

Mid–Atlantic Tilled Land (1991)
Total: 3,980,223 Hectares

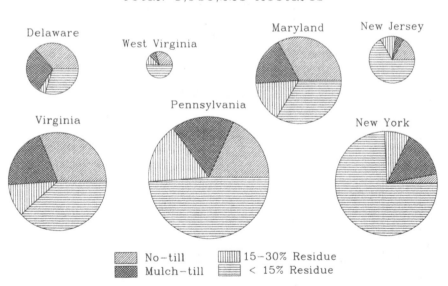

Figure 4. New England (a) and Mid-Atlantic (b) tilled land in 1991. The circle size represents the total tilled area in that state. (Data from NACD-CTIC, 1991.)

determining the influence tillage systems have on the primary microclimatic, chemical, and physical controls regulating these processes. In this way, the peculiarities and potential for success of a given crop residue management-soil-plant system might be more accurately assessed. These controls and processes are often similar for residues and waste products from off-farm sources. The following is a brief description of some of these controls.

A. Soil Water

Soil water is generally increased throughout the growing season for tillage systems leaving 50% or more of the soil surface covered with residues after planting (Griffith et al., 1986). In droughty soils of the Northeast, this may improve yield potential. However, the additional water may reduce yields on soils with poor drainage (e.g. Aquepts, Aquults, Aquods, Aqualfs).

Higher soil water contents are often reported for NT than moldboard plowing (MP) (Blevins et al., 1971; Hill and Blevins, 1973; Linn and Doran, 1984; Griffith et al., 1986; Munawar et al., 1990). In Kentucky, Munawar et al. (1990) found soil water contents under chisel plow and disk tillage treatments to be intermediate between NT and MP.

Soil water content can dramatically impact crop residue decomposition and nutrient cycling (Stanford and Epstein, 1974; Cassman and Munns, 1980; Sommers et al., 1981; Doel et al., 1990). Optimal soil water potential for residue decomposition generally occurs at approximately -0.03 MPa (Sommers et al., 1981).

Water was considered the primary factor influencing microbial populations in NT vs. MP soils (Doran, 1980). Higher soil water content may result in greater denitrification rates in NT compared to MP (Rice and Smith, 1982; Doran, 1980). However, this is only an indirect function of tillage system. The fact that denitrification is related to soil water via residue cover and not directly to tillage system was demonstrated in Kentucky on a Tilsit silt loam (Typic Fragiudults) (Rice and Smith, 1982). A wheat (*Triticum aestivum* L.) winter cover crop had grown poorly at that site, resulting in only a minimal mulch cover. Although this soil was under NT management, lower soil water content and less denitrification were observed on the NT soil compared to its MP counterpart. Such observations are the exception, yet they are valuable for identifying pathways of factor-process response.

Crop residue N mineralization rates are reduced at low soil water contents (Doel et al., 1990). Higher potentially mineralizable N contents have been found in the surface 75 mm of NT soils apparently owing to the higher microbial biomass in that depth (Doran, 1980). However, the potential rate of mineralization may actually be higher under MP due to the larger aerobic microbial population at the 75 to 150 mm depth of MP soils (Doran, 1980).

Nitrification rate of NH_4^+ fertilizer has been proposed to be higher for NT than MP due to more favorable moisture contents under NT (Rice and Smith,

1983). However, in the absence of NH_4^+ fertilizer, Rice and Smith (1983) hypothesized lower total nitrification under NT due to lower mineralization rates and less favorable spatial distribution of nitrifier substrate. Doran (1980) reported higher counts of NH_4^+ and NO_2^- oxidizers (aerobic, autotrophic nitrifiers) in the surface 75 mm of NT compared to MP at four of seven study sites across the U.S.

Higher soil moisture may result in increased plant uptake of those nutrients which move primarily by diffusion. Thomas (1986) hypothesized greater P diffusion under NT not only due to higher soil water content but also to enhanced P availability related to reduced Al solubility in the higher organic matter NT soils.

These studies indicate crop residue management strategies can have a substantial impact on soil water. Thus a soil's internal drainage should be evaluated as a precursor to choosing a crop residue management system. This is easily done by digging a soil pit and describing the soil's morphology. Potential evapotranspiration and nutrient and drainage requirements of alternative crops should also be considered.

B. Soil Temperature

Residue management induced differences in soil temperature are important because temperature influences germination, plant growth rate, nutrient uptake, and several chemical and microbial processes regulating nutrient availability in soils. Lower soil temperatures are a major concern for NT crop production in the northern sector of the northeast region. Griffith et al. (1986) reports till-planting in an elevated ridge to be more popular than NT above 42°N latitude in the northern corn belt.

Surface crop residues generally have a higher solar reflectivity and a lower thermal conductivity than soil (Johnson and Lowery, 1985). Tillage induced differences in soil bulk density and water content influence soil thermal conductivity and heat capacity (Wierenga et al., 1982). Tillage can also affect soil temperature by changing the soil surface configuration (Radke, 1982; Gupta et al., 1990; Benjamin et al., 1990).

Soil temperatures in conservation tillage systems are consequently lower than for MP (Burrows and Larson, 1962; Griffith et al., 1986; Al-Darby and Lowery, 1987; Fortin and Pierce, 1990). Daily minimum soil temperatures are generally similar between MP and NT, with differences resulting from the higher daily maximum temperatures observed under MP (Burrows and Larson, 1962; Fortin and Pierce, 1990).

Seed depth (5 cm) soil temperature and corn emergence were studied in three conservation tillage systems (till-plant (TP), chisel (CH), and NT) and compared to MP in Wisconsin (Al-Darby and Lowery, 1987). Soil temperatures decreased in the order MP > TP > CH > NT, and corn emergence rate followed the order MP > TP = CH > NT. Griffith et al. (1973) compared eight tillage-planting systems in northern, eastern, and southern Indiana for four years. Till-planting was the only system deemed successful regardless of latitude, and this was only for well drained soils.

Plant height, leaf area, and dry matter were all highly correlated with soil thermal units (degree days) in the Wisconsin study of Al-Darby and Lowery (1987). Corn seedling emergence was also related to seed zone soil thermal units for two soils from Minnesota and one soil from Iowa studied by Schneider and Gupta (1985).

Some recent research has focused on assessing the utility of soil thermal units for predicting crop residue decomposition and N mineralization. This approach is appealing because the degree day unit is not only familiar to growers, but it may also be used in common to describe both N mineralization and plant growth, thereby providing information on the quantity and time to apply additional N (Honeycutt et al., 1993a).

This approach was investigated for a Maine soil (Typic Haplorthod) amended with papermill sludge (Honeycutt et al., 1988). Thermal units were shown to describe sludge decomposition regardless of temperature effects and changing substrate lability over time. Commencement of net N mineralization was predicted well with thermal units, with values ranging from 600 to 684 degree days at constant laboratory temperatures of 10, 15, 20, and 25°C. Net N mineralization from field-applied sludge began at approximately 649 degree days.

Honeycutt and Potaro (1990) field tested the application of thermal units for predicting net N mineralization from corn (*Zea mays* L. cv. "King 1113") and lupin (*Lupinus albus* L. cv. "Ultra") residues. These materials were added to soil microplots in Maine during May, June, and July to provide a range of climatic conditions. Commencement of net N mineralization in both corn and lupin residue treatments occurred 119, 99, and 317 days after application which corresponded to 2346, 1990, and 2360 degree days after application, respectively. The observation that net N mineralization from corn and lupin residues applied in July did not begin until late the following May is of particular importance. This indicated that thermal units are valid for predicting commencement of crop residue net N mineralization despite harsh environmental conditions and wide temperature variations to which these residues and soils were subjected.

The influence of soil water content on predicting crop residue N mineralization with thermal units was investigated by Doel et al. (1990). The soil and lupin residue used in that study were the same as those used by Honeycutt and Potaro (1990). Thermal units adequately predicted net N mineralization at -0.03 and -0.01 MPa but not at -0.30 MPa. The thermal unit requirement for onset of lupin residue net N mineralization was similar to that reported in the complementary field study of Honeycutt and Potaro (1990).

Thermal units have also been investigated for predicting nitrate formation across a wide range of soils and fertilizers (Honeycutt et al., 1991). That analysis indicated nitrate concentration X thermal unit relations may be relatively soil-specific as the relationship is apparently influenced by soil texture, pH, and climate.

Honeycutt et al. (1993b) also investigated the effects of residue quality (chemical composition), loading rate, and their interaction on predicting hairy vetch (*Vicia villosa* Roth) residue N mineralization with thermal units. Residue loading rate did not affect residue C or N mineralization. However, residue quality exhibited a dramatic effect on vetch C and N mineralization. Approximately 35 % of the added C had mineralized 30 days after application of a fall-harvested vetch and only about 17 % of added C had mineralized 30 days after application of a spring-harvested vetch that had winter-killed. It is particularly noteworthy that the C/N ratios of these two residues were nearly identical (11.5 and 12.0, respectively). Both lignin and hemicellulose contents increased almost two-fold from the fall to the spring residue collection dates. Consequently, the frequently espoused usage of C/N ratio as a panacea for characterizing residue decomposability should be critically re-examined. Residue C/N ratio does often provide a useful index of residue decomposability; however, other factors may override this relationship as reviewed by Honeycutt et al. (1993b).

Wolf and Rogowski (1991) recently demonstrated how soil heat-flux distributions in Pennsylvania could be used to calculate soil thermal units, thereby delineating land-use management zones on both farm and watershed scales. Douglas and Rickman (1992) recently found thermal units based on air temperatures could also be used to describe decomposition of a relatively wide range of crop residues and climates. This approach may have particular application in areas with limited soil temperature information.

Thus choice of residue and nutrient management systems should consider latitude and microclimatic parameters such as aspect and slope position as these will influence soil thermal properties. Conservation tillage systems generally result in lower soil heat load compared to MP. However, increased emphasis on soil temperatures rather than air temperatures may one day allow linking crop nutrient uptake with crop residue nutrient mineralization. Lastly, RT appears to have been only scarcely tested in the Northeast. This system would appear to overcome the longer soil warm-up limitations of NT in the northern sector of the region (e.g. > 42°N latitude). The potential of RT in the Northeast should be thoroughly examined.

C. Soil pH

Another factor affected by residue management which in turn influences nutrient cycling is soil pH. It was previously mentioned that Doran (1980) reported higher counts of aerobic, autotrophic nitrifiers in the surface 75 mm of NT compared to MP at four of seven study sites across the U.S. Soil pH under NT was 0.4 units lower than the corresponding MP soil at the other three locations. Thus the higher pH under MP may have resulted in greater nitrifier populations under MP at these three sites. Lower soil pH under NT has also been reported in Kentucky for a Maury silt loam (Typic Paleudalfs) (Blevins et al., 1977) and in Ohio for both a Hoytville silty clay loam (Mollic Ochraqualfs) and a Wooster silt loam (Typic Fragiudalfs) (Dick, 1983). Soil pH under conservation tillage should probably be monitored even more closely than under MP, as pH can dramatically influence nutrient transformations from both organic and inorganic sources. The selection of cropping system, amount and type of fertilizers, soil drainage, and soil buffer capacity will influence the impact of tillage system on soil pH.

D. Spatial Distribution

A seemingly important question for understanding crop residue decomposition and nutrient cycling in conservation tillage is the role of residue placement on these processes. Incorporated vs. non-incorporated residue represents a principal difference in agroecosystem structure pertinent to residue decomposition. Non-incorporated residues may be physically separated from the soil, restricting both substrate C accessibility to soil microorganisms and soil inorganic N accessibility to microorganisms inhabiting surface residues.

Noteworthy interactions among residue spatial distribution, soil temperature, and soil water on residue decomposition should also be realized. Slow decomposition of spatially inaccessible residues is conceivable at soil temperature and water conditions considered optimal for microbial activity. Residue placement affects microclimate and microbial populations. Higher populations of fungi are promoted under surface residue conditions than when residue is incorporated (Holland and Coleman, 1987). This increased ratio of fungal to bacterial activity in NT may result in lower soil organic matter losses in NT due to greater substrate use efficiency of fungi and to a proportionally greater C accumulation in the less decomposable fungal biomass (Holland and Coleman, 1987).

Most studies on residue placement effects have been conducted in the western U.S. However, many of the principles may be transferable to the northeast region. Decomposition of surface residues is slower than for incorporated residues. Approximately 50% of a surface residue and 65% of a buried residue of cornstalks had decomposed after 20 weeks in an Iowa field study (Parker, 1962). Winter wheat (*Triticum aestivum* L., cv. "Hyslop") residue losses were 25, 31, and 85% from straw placed above, on, and in the soil, respectively, after 26 months in an Oregon field study (Douglas et al., 1980).

Douglas and Rickman (1992) considered the influence of residue placement to be primarily related to the control of soil water content on residue decomposition. Thus equations for simulating residue decomposition utilized "water coefficients" ranging from 0.2 to 0.3 for surface residues and 0.8 to 1.0 for buried residues. In a Texas greenhouse study, wheat residues mixed with soil and residues covered with a layer of soil decomposed similarly and at a much higher rate than surface residues (Unger and Parker, 1968).

Although decomposition of surface residues may occur slower than for incorporated residues, greater amounts of inorganic P may be leached into the soil from surface residues than incorporated residues. This was shown for several different crop residues in a laboratory study in Oklahoma (Sharpley and Smith, 1989).

Thus, residue placement affects key microclimatic parameters and in some management systems may be paramount in controlling rates of crop residue decomposition and nutrient turnover.

VII. Pros and Cons of Residue Use on Agricultural Land

There are advantages and disadvantages in using residues and waste materials on farm fields. Benefits of using residues as mulches or soil amendments include erosion protection, soil structure enhancement, and nutrient recycling (Logan et al., 1987; Hoffman, 1991). Research has shown that certain residues can assist in controlling weeds and plant diseases (Worsham, 1991; Putnam, 1988). Residues, at times, may have beneficial effects on soil water and soil temperature, although the opposite often occurs in northern climates. Residues also may harbor pests including weed seeds which may result in increased plant stresses. Residues from urban and industrial sources may contain unwanted biological or chemical constituents that would preclude their use on agricultural land. Economics will also determine the feasibility of using off-farm residues.

Cover crops provide many of the advantages of crop residues including runoff and erosion control (Langdale et al., 1991), increased soil tilth (Bruce et al., 1991), and weed and pest control (Worsham, 1991; Bugg, 1991). Living plants may also use soil nutrients that would otherwise be lost to leaching during the fall and spring periods (Radke et al., 1988). In addition, legumes can provide nitrogen for succeeding crops (Radke et al., 1987). Cover crops may use water and nutrients or in other ways compete with the primary crop making cover crops an undesirable alternative, especially in droughty areas. Appropriate plant species and management practices are necessary for successful utilization of cover crops (Hofstetter, 1988; USDA-CSRS, 1992). A comprehensive report on cover crops discussing the above topics is given in Hargrove (1991).

Large numbers of farm animals are raised in the Northeast (Table 7). Animal manures are a good source of nutrients and organic matter. Proper storage, handling, and application are necessary to reduce undesirable odor and nutrient losses via runoff and leaching. Animal manure is also a good material for co-composting with low-nitrogen residue such as straw or leaves.

Processing wastes often contain valuable nutrients and organic matter beneficial to agriculture (Glenn, 1988). Wastes containing large amounts of heavy metals, pest or disease vectors, or toxic chemicals are not suitable for disposal on farm fields. Processing wastes often are deficient in certain nutrients and may need to be combined with other materials to be suitable as mulches or soil amendments. For example, wastes low in N may be mixed with high-N animal manure, or wastes high in N may be mixed with leaves, straw, or peat to provide more suitable materials for spreading on agricultural land (Mathur et al., 1988).

Urban waste consists of large amounts of paper and yard wastes (Figure 3). Sewage effluent and sludge are also available for farm spreading (Hornick, 1988; Elliott, 1986). Paper that is not recycled can be shredded for animal bedding and mulch. Paper is a good bedding material that is considerably cheaper than straw. Paper also can be composted or co-composted with other materials (Gresham et al., 1990). Leaves and grass are often spread directly on the land (Crable, 1991). New Jersey has a large program in leaf mulching and has developed stringent state regulations on landspreading these materials (Dan Kluchinski, Rutgers Cooperative Extension, 1992 personal communication). Farmers can spread leaves on their fields up to 15 cm deep. Twelve percent of the municipalities in New Jersey were participating in this program in 1990. Wind blowing leaves off the field can be a problem. Rutgers University experiments (Dan Kluchinski, 1992 personal communication) have shown fewer pests (especially nematodes) and diseases associated with leaf mulching. More grassy weeds grew because of the increased water content under the leaf mulches. More studies of this type are needed as urban wastes are increasingly applied to agricultural land.

VIII. Composting of Residues and Wastes

Composting is receiving considerable attention as a possible solution to waste disposal. Gardeners have made and used compost for many years. Now some municipalities are considering the construction of large plants to compost urban and industrial wastes (Logsdon, 1989; State of CT DEP, 1989). Leaves, grass clippings, tree trimmings, newspaper, separated garbage, sludge, and manure may be good composting materials. The Rodale Research Center has a large research effort directed towards co-composting leaves, grass clippings, and newspapers with manure to obtain a better nutrient balance for crops (Gresham et al., 1990).

Composting requires proper aeration during the process. Composting plants may use a vessel which is agitated to provide mixing or has forced aeration. In open areas, compost piles can be mixed with earth-moving equipment or run through mixing devices like a stationary manure spreader. Handling, processing, and transporting compost to the field can be costly procedures. Some compost may be saleable to gardeners and homeowners, but probably not at a price that would cover the entire operation. Likewise, research has shown that it would not be economically feasible (with the present cost of commercial fertilizers) for farmers to co-compost waste materials without receiving a tippage fee (Dreyfus, 1990).

Compost can be an excellent soil amendment (Reider et al., 1991). Compost has many of the advantages of crop residues and may be a better material for enhancing soil tilth than straw or stover. Properly processed compost should be devoid of pests and disease-causing organisms. Care will be needed to prevent unwanted chemicals and inert materials in the compost material. State and federal rules and regulations will make large composting plants costly to build and operate. Producers of waste materials may have to shoulder a large portion of the cost associated with waste disposal whether it be through composting, direct application on agricultural land, or by other means.

IX. Summary

Residue management in the Northeast is faced with several challenges that separate this region from the remainder of the U.S. Agriculture, while important, has less impact on the total economy of the Northeast compared to other regions. Conservation tillage and cover crops do play important roles in farming operations in several areas but also have some unique problems. Proper handling, application, and utilization of manure and crop residues are needed to protect soil productivity and prevent environmental damage through erosion, runoff, leaching, and volatilization. Knowledge of tillage-system impacts on key parameters regulating nutrient cycling may provide the information required to develop site- and residue-specific management strategies that are economical and environmentally responsible.

Urban and industrial waste materials will be increasingly applied to agricultural land because of the dense population and industrialization in parts of the Northeast combined with increasing costs of waste disposal. Incineration, disposal in landfills, and ocean dumping are going to be more difficult and costly in the future, and more pressure will be brought to dispose of waste materials on agricultural lands. More research on composting, co-composting, and other techniques is required to assure that application on agricultural land will not adversely affect humans, animals, agriculture, or the environment.

References

Al-Darby, A.M. and B. Lowery. 1987. Seed zone soil temperature and early corn growth with three conservation tillage systems. *Soil Sci. Soc. Am. J.* 51:768-774.

Benjamin, J.G., A.D. Blaylock, H.J. Brown, and R.M. Cruse. 1990. Ridge tillage effects on simulated water and heat transport. *Soil and Tillage Res.* 18:167-180.

Blevins, R.L., D. Cook, S.H. Phillips, and R.E. Phillips. 1971. Influence of no-tillage on soil moisture. *Agron. J.* 63:593-596.

Blevins, R.L., G.W. Thomas, and P.L. Cornelius. 1977. Influence of no-tillage and nitrogen fertilization on certain soil properties after 5 years of continuous corn. *Agron. J.* 69:383-386.

Bruce, R.R., P.F. Hendrix, and G.W. Langdale. 1991. Role of cover crops in recovery and maintenance of soil productivity. p. 109-115 In: W.L. Hargrove (ed.) Cover crops for clean water: the proceedings of an international conference. Soil and Water Conserv. Soc., Ankeny, IA.

Bugg, R.L. 1991. Cover crops and control of arthropod pests of agriculture. p. 157-163. In: W.L. Hargrove (ed.) Cover crops for clean water: the proceedings of an international conference. Soil and Water Conserv. Soc., Ankeny, IA.

Burrows, W.C. and W.E. Larson. 1962. Effect of amount of mulch on soil temperature and early growth of corn. *Agron. J.* 54:19-23.

Cassman, K.G. and D.N. Munns. 1980. Nitrogen mineralization as affected by soil moisture, temperature, and depth. *Soil Sci. Soc. Am. J.* 44:1233-1237.

Ciolkosz, E.J., T.W. Gardner, and J.C. Sencindiver. 1984. Geology, physiography, vegetation and climate. p. 2-14. In: R.L. Cunningham and E.J. Ciolkosz (eds.) Soils of the Northeastern United States. Pennsylvania State Univ. Agric. Exp. Sta. Bull. 848.

Conservation Technology Information Center. 1991. CEDAR user's guide for CTIC electronic data access and retrieval. NACD's Conserv. Technology Information Center, West Lafayette, IN. 25 pp.

Cox, W.J., D.J. Otis, H.M. van Es, F.B. Gaffney, D.P. Snyder, K.R. Reynolds, and M. van der Grinten. 1992. Feasibility of no-tillage and ridge tillage systems in the Northeastern USA. *J. Prod. Agric.* 5:111-117.

Cox, W.J., R.W. Zobel, H.M. van Es, and D.J. Otis. 1990. Growth development and yield of maize under three tillage systems in the northeastern U.S.A. *Soil and Tillage Res.* 18:295-310.

Crable, S. 1991. Agricultural utilization of yard waste. *BioCycle* 32(8): 54-57.

Dick, W.A. 1983. Organic carbon, nitrogen, and phosphorus concentrations and pH in soil profiles as affected by tillage intensity. *Soil Sci. Soc. Am. J.* 47:102-107.

Doel, D.S., C.W. Honeycutt, and W.A. Halteman. 1990. Soil water effects on the use of heat units to predict crop residue carbon and nitrogen mineralization. *Biol. Fertil. Soils* 10:102-106.

Doran, J.W. 1980. Soil microbial and biochemical changes associated with reduced tillage. Soil Sci. Soc. Am. J. 44:765-771.

Douglas, C.L., Jr., R.R. Allmaras, P.E. Rasmussen, R.E. Ramig, and N.C. Roager, Jr. 1980. Wheat straw composition and placement effects on decomposition in dryland agriculture of the Pacific Northwest. *Soil Sci. Soc. Am. J.* 44:833-837.

Douglas, C.L., Jr. and R.W. Rickman. 1992. Estimating crop residue decomposition from air temperature, initial nitrogen content, and residue placement. *Soil Sci. Soc. Am. J.* 56:272-278.

Dreyfus, D. 1990. *Feasibility of on-farm composting.* Rodale Research Center RU-90/2. Rodale Institute, Emmaus, PA. 32 pp.

Elliott, H.A. 1986. Land application of municipal sewage sludge. *J. Soil and Water Conserv.* 41:5-10.

Flint, R.F. 1971. *Glacial and quaternary geology.* John Wiley and Sons, New York, NY. 892 pp.

Fortin, M.-C. and F.J. Pierce. 1990. Developmental and growth effects of crop residues on corn. *Agron. J.* 82:710-715.

Foss, J.E., F.R. Moormann, and S. Rieger. 1983. Inceptisols. In: L.P. Wilding, N.E. Smeck, and G.F. Hall (eds.) *Pedogenesis and soil taxonomy. II. The soil orders.* Elsevier, Amsterdam.

Gile, L.H., Jr. 1958. Fragipan and water table relationships of some Brown Podzolic and Low Humic-Gley soils. *Soil Sci. Soc. Am. Proc.* 22:560-565.

Glenn, R.C. 1988. *Recycling waste potatoes on the farm.* Maine Agric. Exp. Sta. Misc. Rpt. 318. Univ. of Maine, Orono, ME. 5 pp.

Gresham, C.W., R.R. Janke, and J. Moyer. 1990. *Composting of poultry litter, leaves and newspaper.* Rodale Research Center RU-90/1. Rodale Institute, Emmaus, PA. 37 pp.

Griffith, D.R., J.V. Mannering, and J.E. Box. 1986. Soil and moisture management with reduced tillage. p. 19-57. In M.A. Sprague and G.B. Triplett (eds.) *No-tillage and surface-tillage agriculture.* John Wiley and Sons, New York.

Griffith, D.R., J.V. Mannering, H.M. Galloway, S.D. Parsons, and C.B. Richey. 1973. Effect of eight tillage-planting systems on soil temperature, percent stand, plant growth, and yield of corn on five Indiana soils. *Agron. J.* 65:321-326.

Gupta, S.C., J.K. Radke, J.B. Swan, and J.F. Moncrief. 1990. Predicting soil temperatures under a ridge-furrow system in the U.S. Corn Belt. *Soil and Tillage Res.* 18:145-165.

Hanna, W.E., L.A. Daugherty, and R.W. Arnold. 1975. Soil-geomorphic relationships in a first-order valley in central New York. *Soil Sci. Soc. Am. Proc.* 39:716-722.

Hargrove, W.L. (ed.). 1991. *Cover crops for clean water: the proceedings of an international conference.* April 9-11, 1991, Jackson, TN. Soil and Water Conserv. Soc., Ankeny, IA. 198 pp.

Hill, J.D. and R.L. Blevins. 1973. Quantitative soil moisture use in corn grown under conventional and no-tillage methods. *Agron. J.* 65:945-949.

Hoffman, L. 1991. Crop residue management in the Northeast. In: *Crop residue management for conservation.* Proceedings of a national conference. August 8-9, 1991, Lexington, KY. Soil and Water Conserv. Soc., Ankeny, IA. 49 pp.

Hofstetter, R. 1988. Cover Crop Guide. *The New Farm* 10(1):17-22, 27-28, 30-31. Rodale Press, Emmaus, PA.

Holland, E.A. and D.C. Coleman. 1987. Litter placement effects on microbial and organic matter dynamics in an agroecosystem. *Ecology* 68:425-433.

Honeycutt, C.W. and L.J. Potaro. 1990. Field evaluation of heat units for predicting crop residue carbon and nitrogen mineralization. *Plant and Soil* 125:213-220.

Honeycutt, C.W., W.M. Clapham, and S.S. Leach. 1993a. A functional approach to efficient nitrogen use in crop production. *Ecol. Modelling* (in press).

Honeycutt, C.W., L.J. Potaro, K.L. Avila, and W.A. Halteman. 1993b. Residue quality, loading rate, and soil temperature relations with hairy vetch (*Vicia villosa* Roth) residue carbon, nitrogen and phosphorus mineralization. *Biol. Agric. and Hort.* 9:181-199.

Honeycutt, C.W., L.J. Potaro, and W.A. Halteman. 1991. Predicting nitrate formation from soil, fertilizer, crop residue, and sludge with thermal units. *J. Environ. Qual.* 20:850-856.

Honeycutt, C.W., L.M. Zibilske, and W.M. Clapham. 1988. Heat units for describing carbon mineralization and predicting net nitrogen mineralization. *Soil Sci. Soc. Am. J.* 52:1346-1350.

Hornick, S.B. 1988. Use of organic amendments to increase the productivity of sand and gravel spoils: Effect on yield and composition of sweet corn. *Am. J. Alternative Agric.* 3(4):156-162.

Inglett, G.E. (ed.). 1973. *Symposium: Processing agricultural and municipal wastes.* AVI Pub. Co., Inc. Westport, CT. 221 pp.

Johnson, M.D. and B. Lowery. 1985. Effect of three conservation tillage practices on soil temperature and thermal properties. *Soil Sci. Soc. Am. J.* 49:1547-1552.

Langdale, G.W., R.L. Blevins, D.L. Karlen, D.K. McCool, M.A. Nearing, E.L. Skidmore, A.W. Thomas, D.D. Tyler, and J.R. Williams. 1991. Cover crop effects on soil erosion by wind and water. p. 15-22. In : W.L. Hargrove (ed.) Cover crops for clean water. Proceedings of an international conference. Soil and Water Conserv. Soc., Ankeny, IA.

Linn, D.M. and J.W. Doran. 1984. Aerobic and anaerobic microbial populations in no-till and plowed soils. *Soil Sci. Soc. Am. J.* 48:794-799.

Logan, T.J., J.M. Davidson, J.L.Baker, and M.R. Overcash. 1987. *Effects of conservation tillage on groundwater quality: nitrates and pesticides.* Lewis Pub., Chelsea, MI. 292 pp.

Logsdon, G. 1989. A business-like approach to leaf composting. *BioCycle* 30(3):22-24.

Mathur, S.P., J.-Y. Daigle, J.L. Brooks, M. Levesque, and J. Arsenault. 1988. Composting seafood wastes. *BioCycle* 29(8):44-49.

Miller, F.T. and L.A. Quandt. 1984. Soil classification. p. 15-22. In: R.L. Cunningham and E.J. Ciolkosz (eds.) Soils of the Northeastern United States. Pennsylvania State Univ. Agric. Exp. Sta. Bull. 848.

Morrison, F.B. 1957. *Feeds and feeding.* 22nd ed. Morrison Pub. Co., Ithaca, NY. 1165 pp.

Munawar, A., R.L. Blevins, W.W. Frye, and M.R. Saul. 1990. Tillage and cover crop management for soil water conservation. *Agron. J.* 82:773-777.

National Association of Conservation Districts, CTIC. 1991. *National survey of conservation tillage practices* (Including other tillage types). NACD-CTIC. West Lafayette, IN.

Parker, D.T. 1962. Decomposition in the field of buried and surface-applied cornstalk residue. *Soil Sci. Soc. Am. Proc.* 26:559-562.

PC GLOBE, Inc. 1990. *Statistical database for Northeastern states.* PC USA Version 2.0 CD-ROM. Tempe, AZ.

Putnam, A.R. 1988. Allelopathy: Problems and opportunities in weed management. p. 77-88. In: M.A. Altieri and M. Liebman (eds.) *Weed management in agroecosystems: Ecological approaches.* CRC Press, Inc., Boca Raton, FL.

Radke, J.K. 1982. Managing early season soil temperatures in the northern corn belt using configured soil surfaces and mulches. *Soil Sci. Soc. Am. J.* 46:1067-1071.

Radke, J.K., R.W. Andrews, R.R. Janke, and S.E. Peters. 1988. Low-input cropping systems and efficiency of water and nitrogen use. p. 193-218. In: W.L. Hargrove (ed.) Cropping strategies for efficient use of water and nitrogen. ASA Spec. Pub. No. 51. Am. Soc. Agron., Madison, WI.

Radke, J.K., W.C. Liebhardt, R.R. Janke, and S.E. Peters. 1987. Legumes in crop rotations as an internal nitrogen source for corn. p. 56-57. In: J.F. Power (ed.) The role of legumes in conservation tillage systems. Soil Conserv. Soc. Am., Ankeny, IA.

Reider, C., R.R. Janke, and J. Moyer. 1991. Compost utilization for field crop production. Tech. Rpt. RU-91/1. Rodale Institute, Emmaus, PA. 76 pp.

Rice, C.W. and M.S. Smith. 1982. Denitrification in no-till and plowed soils. *Soil Sci. Soc. Am. J.* 46:1168-1173.

Rice, C.W. and M.S. Smith. 1983. Nitrification of fertilizer and mineralized ammonium in no-till and plowed soil. *Soil Sci. Soc. Am. J.* 47:1125-1129.

Schneider, E.C. and S.C. Gupta. 1985. Corn emergence as influenced by soil temperature, matric potential, and aggregate size distribution. *Soil Sci. Soc. Am. J.* 49:415-422.

Seekins, B. and L. Mattei. 1990. *Update: Usable waste products for the farm.* Maine Dept. of Agric., Food, and Rural Resources. Augusta, ME. 103 pp.

Sharpley, A.N. and S.J. Smith. 1989. Mineralization and leaching of phosphorus from soil incubated with surface-applied and incorporated crop residue. *J. Environ. Qual.* 18:101-105.

Smith, H. 1984. General soil map of the Northeastern United States. In: Soils of the Northeastern United States, R.L. Cunningham and E.J. Ciolkosz (eds.). Pennsylvania State Univ. Agric. Exp. Sta. Bull. 848. p. 47.

Soil Survey Staff. 1975. Soil Taxonomy: A basic system of soil classification for making and interpreting soil surveys. USDA-SCS Agric. Handb. 436. 754 p. U.S. Gov. Print. Office, Washington, DC.

Sommers, L.E., C.M. Gilmour, R.E. Wildung, and S.M. Beck. 1981. The effect of water potential on decomposition processes in soils. p. 97-117 In: J.F. Parr, W.R. Gardner, and L.F. Elliott (eds.). Water potential relations in soil microbiology, Soil Sci. Soc. Am. Spec. Pub. No. 9.

Stanford, G. and E. Epstein. 1974. Nitrogen mineralization-water relations in soils. *Soil Sci. Soc. Am. Proc.* 38:103-107.

State of Connecticut Department of Environmental Protection. 1989. Leaf Composting: A guide for municipalities. Prepared by the Univ. of Connecticut Cooperative Ext. Serv., Hartford, CT. 39 pp.

Thomas, G.W. 1986. Mineral nutrition and fertilizer placement. p. 93-116. In: M.A. Sprague and G.B. Triplett (eds.). No-tillage and surface-tillage agriculture. John Wiley and Sons, New York.

Unger, P.W. and J.J. Parker, Jr. 1968. Residue placement effects on decomposition, evaporation, and soil moisture distribution. *Agron. J.* 60:469-472.

U.S. Department of Agriculture, CSRS. 1992. *Managing cover crops profitably.* Produced and edited by the Rodale Institute, Emanus, PA. 114 pp.

U.S. Department of Agriculture, ERS. 1991. *Economic indicators of the farm sector. State Financial Summary, 1989.* Economic Res. Serv. ECIFS 9-3. Washington, DC. 238 pp.

U.S. Department of Commerce. 1968. *Climatic atlas of the United States.* Environmental Science Services Administration, Washington, DC.

U.S. Department of Commerce. 1989. *Census of agriculture (1987).* Geographic area series. Superintendent of Documents, U.S. Gov. Print. Office, Washington, DC.

Wierenga, P.J., D.R. Nielsen, R. Horton, and B. Kies. 1982. Tillage effects on soil temperature and thermal conductivity. p. 69-90. In: P.W. Unger and D.M. Van Doren, Jr. (eds.). Am. Soc. Agron. Spec. Pub. No. 44.

Wolf, J.K. and A.S. Rogowski. 1991. Spatial distribution of soil heat flux and growing degree days. *Soil Sci. Soc. Am. J.* 55:647-657.

Worsham, A.D. 1991. Role of cover crops in weed management and water quality. p. 145-163. In: W.L. Hargrove (ed.) Cover crops for clean water. Proceedings of an international conference. Soil and Water Conserv. Soc., Ankeny, IA.

Yeck, R.D., R.V. Rourke, and R.J. Bartlett. 1984. Spodosols. p. 35-36. In: R.L. Cunningham and E.J. Ciolkosz (eds.) Soils of the Northeastern United States. Pennsylvania State Univ. Agric. Exp. Sta. Bull. 848.

Concepts of Residue Management: Infiltration, Runoff, and Erosion

G.W. Langdale, E.E. Alberts, R.R. Bruce,
W.M. Edwards, and K.C. McGregor

I. Introduction

Crop residues are an important renewable resource that can be managed to conserve nonrenewable soil and water resources and sustain crop production. Soil erosion hazards (water) are usually associated with udic climates of the USA (Soil Survey Staff, 1975). Therefore, management of crop residues on the soil surface is essential to protect soil and water resources.

Managing crop residues at or near the soil surface has been a technological struggle for centuries. European settlers in North America did not possess the conservation expertise to farm highly erodible soils (Bennett, 1947; Trimble, 1974). Beginning in the 1700s, distinguished citizen-farmers of the eastern USA, such as Franklin, Washington, Jefferson, Madison and Ruffin, recognized soil erodibility problems (Bennett, 1939, 1947; Ruffin, 1832). They apparently advocated conservation practices such as contouring and crop rotations which included legume cover crops that controlled some soil erosion. Farmers in Georgia and the Carolinas used "hillside ditching," a forerunner of modern terracing, as early as 1830 to slow soil erosion (Bennett, 1947). Some leading farmers adopted available soil and water conservation technology, but accelerated soil erosion in the U.S. was widespread during the 1800s and early 1900s. For example, a soil survey report completed for Fairfield County, South Carolina in 1911 showed that almost 200,000 ha or 28% of the county's land resource had been so damaged by erosion that the land had no further practical value for immediate crop production (Carr, 1911). Trimble's (1974) historic soil erosion synthesis of the Southern Piedmont suggested that the period of greatest erosive land use occurred between 1860 and 1920. This accelerated soil erosion was attributed to an agricultural system dominated by farm tenancy. These farmers were prone to use poor conservation practices. In the southeastern U.S., poor conservation practices were most often associated with conventionally tilled cotton (*Gossypium hirsutum* L.).

Conservation tillage technology development for cotton has been slow and difficult to develop. This technology has emerged only within the last 5 years, whereas similar technology for other agronomic row crops was adopted 10 to 15 years earlier.

II. Early Crop Residue Management Research

Early research attempts to manage crop residues near the soil surface with inadequate technology were often unsuccessful, and therefore perpetuated poor soil and water conservation practices. Continued erosive farming habits stimulated Bennett's passionate soil conservation leadership that led to the first U.S. legislative action, mandating research for control of soil erosion (Bennett 1939, 1947). This legislation was authorized by the 1928 Buchanan Amendment to the Agricultural Appropriation Bill (Buchanan, 1928). Ten soil erosion experiment stations were funded by the Buchanan Amendment. Wischmeier and Smith (1965) used soil and water conservation data collected in the 1930s and 1940s at these and other soil and water conservation research centers to derive the Universal Soil Loss Equation (USLE). The most sensitive component of the USLE is the residue management component or C-factor.

Poor technology limited early crop residue management; therefore, most research successes were achieved through crop rotations that included a meadow or crude mulch tillage procedures (McCalla and Army, 1961; Sojka et al., 1984 and Williams et al., 1981). Many of the mulch tillage successes included cool season green manuring crops. Excellent examples of this research are cited in publications from Clemson, South Carolina (Beale et al., 1965), Coshocton, Ohio (Edwards and Owens, 1991) and Kingdom City, Missouri (Burwell and Kramer, 1983). Adoption of these technologies was low and ephemeral because management of crop residues remained a struggle.

The ability to successfully manage large quantities of crop residues near the soil surface was accomplished with research conducted in the 1970s and 1980s (Langdale et al., 1991; Unger et al., 1988). The breakthroughs were achieved only with large industry and state/federal government financial support.

Figure 1. Gauged research sites with long-term conservation tillage histories.

III. Long-Term Crop Residue Management for Soil Erosion Control

Some of the largest data gaps in U.S. agricultural research literature are attributed to the dominance of short-term crop residue management systems. Crop residue management research with long-term soil and water conservation objectives is presently near negligible. Only four runoff gauged research sites with data sets of two decades or more exist (Figure 1). The scarcity of these data sets is often attributed to funding cost of long-term research. Watkinsville, Georgia and Coshocton, Ohio research locations are associated with the eastern Ultisol/Alfisol region of the U.S. Holly Springs, Mississippi and Kingdom City, Missouri research locations are associated with Alfisols on the Mississippi drainage basin. Dominating soil taxonomy subgroups are Typic Kandhapludults, Aquic/Aquultic Hapludalfs, Typic Fragiudalfs, and Udollic Ochraqualfs for Georgia, Ohio, Mississippi and Missouri, respectively. Major Land Resource Areas (MLRA) are Southern Piedmont (#136), Allegheny Plateau (#124 and 126), Southern Mississippi Valley Silty Upland (#134), and Central Claypan Area (#113).

Soil and water conservation research was initiated or in progress during the early 1940s at all four of these research locations. The rationale for this early soil and water conservation research was oriented towards the soil erosion hazard of each Major Land Resource Area which these research centers serve.

Table 1. Infiltration rate on severely eroded Kandhapludult soil in Georgia - one hour of simulated rainfall at 60 mm⁻¹ - following 5 and 8 years of cropping

Surface residue	-------CTG^a-------		---NT (CL+G)^b---	
	1988	1991	1988	1991
	--------------------mm h⁻¹--------------------			
With residue	34.5	21/5	46.0	23.2
Residues removed	21.1	13.8	41.6	16.2

^{a}5 years of conventional tillage grain sorghum (1988) plus 3 years of conventional tillage soybean (1991).
^{b}5 years of no-tillage grain sorghum into crimson clover residues (1988) plus 3 years of conventional tillage soybean (1991).
(From West et al., 1991.)

Both prevailing conventional tillage and meadow-based crop rotation practices precede a significant history of conservation tillage. Crude mulch tillage methods were often incorporated into the crop rotation treatments. Improved conservation tillage technologies began to emerge during the 1970s (Unger et al., 1988). These research approaches were subsequently introduced on gauged sites at all four locations. These data sets are the best candidates for describing effects of newly developed conservation managed crop residues on infiltration, runoff, and soil erosion herein of the nation's best udic land resources.

A. Effect of Crop Residues on Infiltration

1. Ultisols of the Southern Piedmont

Infiltration on eroded, conventional tilled, Typic Kandhapludults of the Southern Piedmont is usually inadequate to support warm season crop growth (Bruce et al., 1990c). Because of improved soil structure and biological activity associated with long-term conservation tillage (Bruce et al., 1990b; Langdale et al., 1992a; Radcliffe et al., 1988), infiltration is significantly improved on these soils (West et al., 1991). On eroded Kandhapludults, returning 30 to 60 Mg ha⁻¹ of crop residues to the soil surface over a 3 to 5 year period is required to significantly increase soil carbon and biological activity and subsequently water stable aggregates (Langdale et al., 1984, 1990, 1992b; Bruce et al., 1990a, 1990b). This is demonstrated best with infiltration measurements made by West et al. (1991) following five years of tillage described by Langdale et al. (1992b). The 1988 values (Table 1) include a five-year conservation tillage aggrading treatment. These infiltration measurements, accomplished in spring of 1988, were associated with annual crop residue production of 5.2 Mg ha⁻¹ yr⁻¹ for the conventional tilled grain sorghum (*Sorghum bicolor* L. Moench)–CTG treatment and 11.6 Mg ha^{-1} yr^{-1} for conservation tillage grain sorghum into crimson clover (*Trifolium incarnatum* L.)–NT(CL + G) treatment. When crop residues were managed with conservation tillage, infiltration increased from 34.5 to 46.0 mm h-1. Removing residues becomes more critical on the CTG treatment, reducing infiltration almost 40%.

Three years of conventional tilled soybean (*Glycine max*. L.) followed the 1988 infiltration measurements. The 1991 values (Table 1) show considerable degradation with conventional tillage. However, the three year degrading period does not entirely eliminate the effect of crop residues on infiltration. Soil productivity of eroded Southern Piedmont soils remains low until infiltration of summer rainfall is improved significantly (Langdale et al., 1992a). Infiltration on Typic Kandhapludults of the Southern Piedmont was enhanced by conservation tillage procedures that disrupt restricted subsurface soil layers. Increased infiltration with these conservation tillage procedures was demonstrated in studies conducted by Bruce et al. (1990b) and Langdale et al. (1983). Summary information from these studies is in Tables 2, 3, and 4. The restricting layer occurs between 18 and 24 cm depths or in the upper part of the Bt horizon. In severely eroded soil profiles the restricting layer is much closer to the soil surface. Deep tillage associated with studies shown in Tables 2 and 3 was accomplished with a coulter inrow chisel planter arrangement. Tillage frequencies and associated infiltration values in Table 4 were accomplished with a paraplow on moderately to severely eroded soils.

Deep tillage was required only initially if sufficient quantities (10 to 12 Mg ha^{-1} yr^{-1}) of crop residues are present on the soil surface (Langdale et al., 1990). On severely eroded soil, deep conservation tillage was required more frequently to maintain adequate infiltration. Infiltration on Typic Kandhapludults is most sensitive to surface crop residue management through conservation tillage.

2. Alfisols of the Allegheny Plateau and the Mississippi River Basin

Conservation tillage management of crop residues is extremely important to maintain adequate infiltration on these silt loam soils in this region. Edwards (1982) attributed this primarily to their crusting behavior when conventionally tilled. He also successfully predicted infiltration associated with tillage treatments using stepwise multiple regression equations. A best fit equation required different analyses for crusted and uncrusted soils.

McGregor et al. (1991) measured infiltration on a loess derived silt loam soil underlain with a fragipan. Tillage treatments included no-tillage and conventional tillage, each with a seven year history (Table 5). All plots were primary tilled (disk harrowing) two weeks prior to rainfall simulation. Soil crusting was probably related to the lower infiltration values associated with conventional tillage. Crop residue management with no-till procedures appears equally important to maintaining higher infiltration rates on both Alfisols and Ultisols. Initial infiltration values were 46 and 43 mm h^{-1} for no-tillage and 34 and 31

Table 2. Effect of 8 years summer crop tillage (grain sorghum and soybean rotations) on cumulative infiltration (I) and sorptivity (S) at 2.0 min. after the start of infiltration on a Kandhapludult soil in Georgia

Tillage[a]	Sorptivity	Infiltration
	mm s$^{-1/2}$	mm
Inrow chisel[b]	4.02	44.0
No-tillage	3.61	40.0
Conventional	3.52	40.6

[a]Offset disk harrowed 130 mm deep after fall to plant wheat.
[b]230 mm depth.
(From Bruce et al., 1990.)

Table 3. Effect of tillage, crop cover, initial soil water content and slope length on infiltration of simulated rainfall at a rate of 64 mm h^{-1} on Kandhapludult soil in Georgia

Tillage, cover Initial water content	Slope length, m	
	10.7	21.4
	mm h^{-1}	
In-row chisel in rye mulch		
5.8%	61	52
14.4%	40	21
Soybean canopy (68%) over rye mulch		
8.7%	52	36
14.1%	24	6
Tilled, bare fallow		
12.0%	18	10
14.7%	8	5

(From Langdale et al., 1983.)

Table 4. Steady-state ponded infiltration rate associated with paraplow tillage frequency to a depth of 0.3 m on a Kandhapludult soil at Watkinsville, Georgia

Years following paraplowing	Soil erosion	
	Moderate	Severe
1	100	72
2	66	19
3	17	11

(Unpublished data by D.E. Radcliffe, G.W. Langdale, and R.R. Bruce, Watkinsville, GA 30677)

Table 5. Infiltration rate on Fragiudalf soil following 7 years of tillage history at Holly Springs, Mississippi[a]

Treatment history	Simulated runs[b]		
	Initial	Wet	Very wet
	-----------------------------mm h[-1]----------------------------		
No-till	43	22	12
Conventional till	31	10	6

[a]These rates occurred during rainulator studies following tillage of two diskings for both the conventional-till and no-till areas preceding simulated rainfall.
[b]Simulated rate of 66 mm h[-1] for 60-min. initial, 30-min. wet, 30 min. very wet runs.
(From McGregor et al., 1991.)

mm h[-1] for conventional tillage of respective soil orders (Tables 1 and 5). These data sets may universally demonstrate the value of managing crop residues with conservation tillage.

B. Effect of Crop Residues on Runoff and Soil Erosion

1. Ultisols of the Southern Piedmont

Soil erosion scars in the Southern Piedmont suggest that crop residues have been poorly managed for centuries. Cotton was the dominant cultivated crop for more than 150 years. Research accomplished on continuous conventional tilled cotton runoff plots (1940-60) at the Southern Piedmont Conservation Research Center, Watkinsville, Georgia, mimics soil erosion damage to Southern Piedmont soils (Carreker et al., 1977). Twenty-four percent of the annual rainfall was partitioned into runoff. Accompanying soil losses averaged near 50 Mg ha[-1] yr[-1]. Research accomplished during the 1940s and 1950s on Southern Piedmont lands provided technology to generate sufficient quantities of crop residues only through winter cover crops and warm season sod crop rotations (Carreker et al., 1977; Beale et al., 1965). However, economics and poor conservation tillage technology limited sustained adoption by farmers.

Conservation tillage planting equipment and suitable herbicides emerged on the horizon during the early 1970s (Unger et al., 1988). Data sets represented in a stochastic form in Figures 2 and 3 demonstrate the dramatic effect of this technology on runoff and soil erosion (Langdale et al., 1979; Mills et al., 1986, 1988). The rainfall retention expression shown in curve numbers 2, 3, and 4 (Figure 2) was attributed to the cumulative double crop residue effect as well as deep conservation tillage expressed in curve numbers 3 and 4. These double crop conservation tillage procedures contributed crop residue quantities that

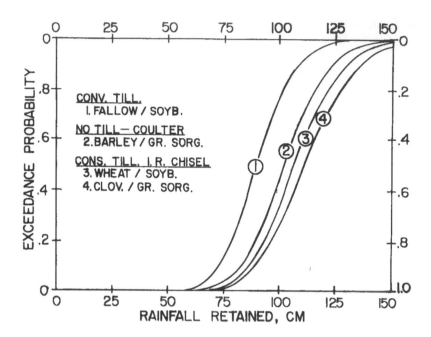

Figure 2. Stochastically derived rainfall retention probabilities of conventional and conservation tilled crop rotations.

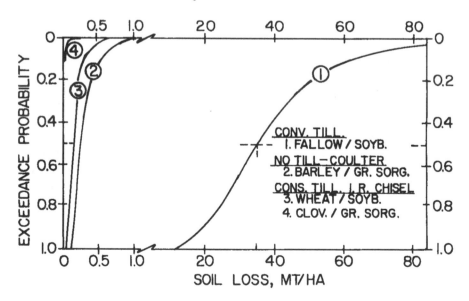

Figure 3. Stochastically derived soil loss probabilities of conventional and conservation crop rotations.

average 10 to 12 Mg ha^{-1} yr^{-1}, compared to less than 5 Mg ha^{-1} yr^{-1} for the monocrop conventional tilled soybean systems. There is a 50% exceedance probability that near 30% less rainfall will be partitioned into runoff when the system approaches equilibrium (curve No. 1 vs. No. 4). Accompanying soil losses (Figure 3), also derived stochastically, decreased from 35 Mg ha^{-1} yr^{-1} to less than 0.5 Mg ha^{-1} yr^{-1} at 50% exceedance probability when conservation tillage was introduced.

Following 17 years of double crop conservation tillage on this watershed, runoff was reduced from 16.2 to 1.8% of the 1260 mm annual average rainfall (Langdale et al., 1992a). Perhaps the most important role of continuous conservation tilled managed crop residues on Hapludult soils is retardation of runoff associated with high erosivity storms (> 100 mm rainfall or an EI > 1000 MJ mm [ha.h]$^{-1}$). On a long-term average, these storms may occur every two years on the Southern Piedmont. Conservation tillage-managed crop residues described above reduced runoff of these storms from 53 to 22% with almost no soil loss over a 17-year period (Langdale et al., 1992a).

2. Alfisols of the Allegheny Plateau and the Mississippi River Basin

Runoff sites at Coshocton, Ohio and Kingdom City, Missouri have been gauged for approximately 50 years. Holly Springs, Mississippi sites were gauged approximately 30 years ago. Soil and water conservation treatments have been compared with poor crop residue management treatments since initiation at all locations.

Modern conservation tillage technology for retaining crop residues at the soil surface was imposed at all sites more than a decade ago (Alberts et al., 1985; Burwell and Kramer, 1983; Edwards and Owens, 1991; McGregor et al., 1991; Edwards, 1982; Mutchler et al., 1985; Mutchler and McDowell, 1990). Summary data sets shown in Tables 6, 7, and 8 are representative of these long-term studies. Runoff from these silt loam Alfisols appears to be greater than on Ultisols. When recent conservation tillage technology is introduced to these runoff sites, runoff may increase. However, this increase in runoff does not appear serious since it never exceeds 10% of the annual rainfall. Soil losses that average near 1.0 Mg ha-1 yr-1 confirm this assessment. More importantly, a crop residue management system for these soils should be employed that develops a C- factor < 0.02 or a SLR < 0.007. Shelton et al. (1983; 1987) obtained similar results on a silt loam Paleudalf in west Tennessee (Table 9).

Generally, rotating a reduced or conventional tillage system on all silt loam soils discussed herein causes soil losses to exceed the soil loss tolerance (T-value) (Mutchler et al., 1985; Mutchler and McDowell, 1990; Edwards and Owens, 1991). In the Coshocton data set (Table 6), most soil losses occurred during corn production years. When Mutchler et al. (1985) grew cotton with conventional tillage following 11 years of no-tillage, soil losses increased from 1.3 to 48.6 Mg ha^{-1} yr^{-1} (Table 8). However, reduced tillage procedures used at

Table 6. Runoff and soil loss results associated with long-term management of Hapludalf soils at Coshocton, Ohio

Treatment	Runoff	Soil loss
	mm (%)	Mg ha^{-1} yr^{-1}
Prevailing practice (1942-1969)[a]	45(4.7)	4.13
Improved practice (1942-1969)[b]	37(3.9)	1.23
Conservation tillage (1984-1990)[c]	63(6.8)	0.50

[a]Conventional tillage with low fertility and straight rows in a corn, wheat, meadow, meadow rotation.
[b]Conventional tillage with high fertility, contour planted in a corn, wheat, meadow, meadow rotation.
[c]Conservation tillages included chisel, paraplow, and no-tilling on two watersheds each.
(From Edwards and Owens, 1991.)

Table 7. Tillage effects on annual C-factor, runoff and soil loss of a clay pan Alfisol, a Udollic Ochraqualf, in Missouri

Tillage methods	C-factor	Runoff (mm)	Soil loss (Mg ha^{-1} yr^{-1})
		Continuous corn	
Conventional	0.08	157	3.38
Field cultivation	0.04	126	1.84
No-tillage	0.01	167	0.51
		Continuous soybean	
Conventional	0.18	144	8.35
Field cultivation	0.08	151	3.29
No-tillage	0.02	161	1.05

(From Alberts et al., 1985.)

Coshocton and Kingdom City from the 1940s through the 1970s controlled soil erosion near 2.0 Mg ha^{-1} yr^{-1} without altering runoff markedly below that of conventional tillage (Table 6, Figures 4 and 5). Unless crop residues are managed with current levels of conservation tillage technology, Alfisol landscapes may suffer serious soil losses. Including meadow rotations with conservation tillage row crops provides an excellent alternate management strategy for these soils.

Table 8. Residual effects of surface managed crop residues on Fragiudult soil in Mississippi

Treatments	Soil loss	Soil loss ratio	Runoff
	Mg ha^{-1} yr^{-1}	%	mm
Conventional tillage cotton after 11 years conventional tillage corn and soybean	91.4	0.411	48
Conventional tillage cotton after 11 years no-till corn and soybean	48.6	0.237	35
Continuous no-till	1.3	0.007	20

(From Mutchler et al., 1985.)

Table 9. Effect of 7 years of soybean/wheat cropping systems on a Paleudalf soil in west Tennessee

Treatment	C-factor	Runoff[a]	Soil loss
		%	Mg ha^{-1}
Continuous till soybean/fallow	0.442	43	7.48
Continuous till soybean/wheat	0.080	41	1.68
No-till soybean/wheat	0.006	46	0.09

[a]Natural and simulated storms averaged 56 mm (R=54).
(From Shelton et al., 1983, 1987.)

IV. Conclusions

Fifty years of conservation research at U.S. research centers succinctly demonstrate the importance of managing crop residues on or near the soil surface. These crop residue strategies generally increase rainfall infiltration and always reduce soil losses. Improved conservation tillage techniques were introduced to a few runoff plots and small watersheds during the 1970s. This minimized crop residue management struggles and dramatically reduced both the C-factor and soil loss ratios to values < 0.02 and 0.007, respectively.

Runoff associated with conservation tillage technology is inconsistent, primarily because of effects of fragi- and clay-pans occurring in soil profiles. However, improved infiltration and soil erosion control were very consistent. Long-term conservation tillage increases infiltration, up to approximately 45 mm h^{-1}, and decreases soil losses to less than 1.0 Mg ha^{-1} yr^{-1}, on both Alfisols and Ultisols of the eastern USA. Runoff is consistently reduced on Ultisols with long-term conservation tillage.

Use of double crop residues to accomplish sound soil and water management is easier to accomplish on udic-thermic Land Resource Areas, where nearly 50%

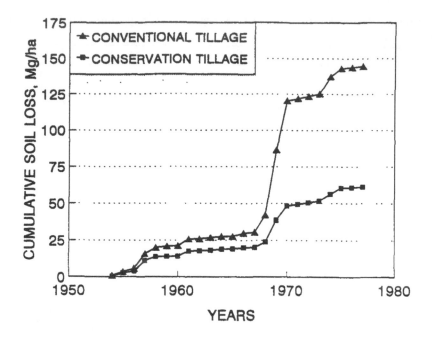

Figure 4. Long-term cumulative soil losses representing conventional and conservation tillages by Kingdom City, MO.

of the total crop residue production (> 10 Mg ha^{-1} yr^{-1}) may be associated with cool season crops. Management of these large quantities of crop residues became successful only recently with improved-aggressive planting equipment. Improved herbicide and crop rotation management also enhances these successes. Management problems may arise in xeric, boreal, and semiarid tropical climate association because of meager crop residue production or rapid decompositions.

Research data sets suggest that best crop residue management successes are achieved with continuous conservation tillage systems. On Ultisols, 3 to 5 years of multiple crop-conservation tillage are required to approach an equilibrium. Longer periods of time may be required to gain equilibrium on other major U.S. soil orders.

Short-term studies may have created the largest research gap in the U.S. Most conspicuous is the absence of the use of modern conservation tillage research in association with long-term sod-based crop rotations. National farm policies (e.g. Conservation Reserve Program) impinge greatly on this research gap and subsequent adaption. Considerable hectarage of sod crops could be grown in udic areas for future conservation tillage. Although water stable aggregation and infiltration are indicative of the improved soil tilth value, soil carbon measurements are inexpensive and serve as a good proxy (Table 10).

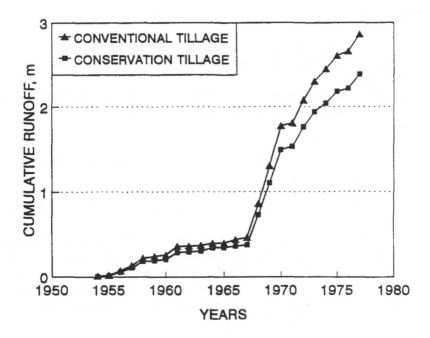

Figure 5. Long-term cumulative runoff representing conventional and conservation tillages by Kingdom City, MO.

Table 10. Effect of tillage on rainfall retention and soil productivity on Kandhapludult soils in Georgia

Carbon[a]	Water stable[a] aggregates	Rainfall retention	Relative grain yield[b]
------	------ %	------	------
	Conventional tillage		
0.70	61	60	50
	Conservation tillage		
2.50	91	80	100

[a]Soil depth = 0.0 to 0.15 m.
[b]Crop yield equivalent to 1.5 and 3.0 Mg ha[-1] soybean yield.
(From Langdale et al., 1992.)

Once these soil properties approach an optimum level, the soil resource is best protected when major storms occur.

References

Alberts, E.E., R.C. Wendt, and R.E. Burwell. 1985. Corn and soybean cropping effects on soil losses and C factors. *Soil Sci. Soc. Am. J.* 49:721-728.

Beale, O. W., G. B. Nutt, and T. C. Peele. 1965. The effect of mulch tillage on runoff, erosion, soil properties and crop yields. *Soil Sci. Soc. Am. Proc.* 19:244-247.

Bennett, H. H. 1939. *Soil Conservation.* McGraw-Hill, New York, NY.

Bennett, H.H. 1947. Development of our national program of soil conservation. *Soil Sci.* 108:259-273.

Bruce, R.R., G.W. Langdale, and A.L. Dillard. 1990a. Tillage and crop rotation effect on characteristics of a sandy surface soil. *Soil Sci. Soc. Am. J.* 54:1744-1747.

Bruce, R. R., G. W. Langdale, and L. T. West. 1990b. Modification of soil characteristics of degraded soil surfaces by biomass inputs and tillage affecting soil water regimes. *TRANS. XIV Cong., Int.Soc. Soil Sci.* VI:4-9.

Bruce, R.R., W.M. Snyder, A.W. White, Jr., A.W. Thomas, and G. W. Langdale. 1990c. Soil variables and interactions affecting prediction of crop yield patterns. *Soil Sci. Soc. Am. J.* 54:494-501.

Buchanan, J.P. 1928. *U.S. Congressional Record.* p. 835. 18 December 1928.

Burwell, R.E. and L.A. Kramer. 1983. Long-term annual runoff and soil loss from conventional and conservation tillage of corn. *J. Soil Water Cons.* 38:315-319.

Carr, M.E. 1911. *Soil survey.* Fairfield County, South Carolina. U.S. Bureau of Soils.

Carreker, J.R., S.R. Wilkinson, A.P. Barnett, and J.E. Box, Jr. 1977. Soil and water management systems for sloping lands. *USDA-ARS-S-160* Agr. Res. Serv., U. S. Dept. Agr., Washington, D. C.

Edwards, W.M. 1982. Predicting tillage effects on infiltration. p. 105-115. In: *Tillage effects on soil physical properties and processes.* Am. Soc. Agron., Madison, WI.

Edwards, W.M. and L.B. Owens. 1991. Large storm effects on total soil erosion. *J. Soil Water Cons.* 46:75-78.

Langdale, G.W., A.P. Barnett, R.A. Leonard, and W.G. Fleming. 1979. Reduction of soil erosion by the no-till system in the Southern Piedmont. *Trans. ASAE* 22:82-86 and 92.

Langdale, G.W., H.F. Perkins, A.P. Barnett, J.C. Reardon, and R.L. Wilson, Jr. 1983. Soil and nutrient runoff losses with in-row chisel-planted soybean. *J. Soil Water Cons.* 38:297-301.

Langdale, G.W., W.L. Hargrove, and J. Giddens. 1984. Residue management in double-crop conservation tillage systems. *Agron. J.* 76:689-694.

Langdale, G.W., R.L. Wilson, and R.R. Bruce. 1990. Cropping frequencies to sustain long-term conservation tillage systems. *Soil Sci. Soc. Am. J.* 54:193-198.

Langdale, G.W., R.L. Blevins, D.L. Karlen, D.K. McCool, M.A. Nearing, E.L. Skidmore, A.W. Thomas, D.D. Tyler, and J.R. Williams. 1991. Cover crop effects on soil erosion by wind and water. p. 15-22. In: W.L. Hargrove (ed.) *Cover crops for clean water.* Soil and Water Conservation Society, Ankeny, IA.

Langdale, G.W., W.C. Mills, and A.W. Thomas. 1992a. Use of conservation tillage to retard erosion effects of large storms. *J. Soil Water Cons.* 47:257-260.

Langdale, G.W., L.T. West, R.R. Bruce, W.P. Miller, and A.W. Thomas. 1992b. Restoration of eroded soil with conservation tillage. *Soil Technology* 5:81-90.

McCalla, T.M. and T.J. Army. 1961. Stubble mulch farming. *Adv. Agron.* 13:125-196.

McGregor, K.C., C.K. Mutchler, and R.F. Cullum. 1991. Soil erosion effects on soybean yield. *Am. Soc. Agric. Engr.* Paper No. 912626. St. Joseph, MO.

Mills, W. C., A. W. Thomas, and G. W. Langdale. 1986. Estimating soil loss probabilities for Southern Piedmont cropping-tillage systems. *Trans. ASAE* 29:948-955.

Mills, W.C., A.W. Thomas, and G.W. Langdale. 1988. Rainfall retention probabilities computed for different cropping tillage systems. *Agric. Water Manage.* 15:61-71.

Mutchler, C.K., L.L. McDowell, and J.D. Greer. 1985. Soil loss from cotton with conservation tillage. *Trans. ASAE* 23:160-163.

Mutchler, C.K. and L.L. McDowell. 1990. Soil loss from cotton with winter cover crops. *Trans. ASAE* 33:432-436.

Radcliffe, D.E., E.W. Tollner, W.L. Hargrove, R.L. Clark, and M.H. Golabi. 1988. Effect of tillage practices on infiltration and soil strength of a Typic Hapludult soil after ten years. *Soil Sci. Soc. Am. J.* 52:798-804.

Ruffin, E. 1832. *An assay on calcareous manures.* (J.D. Sitterson [ed.] 1961). The Belknap Press, Cambridge, MA.

Shelton, C.H. and J.F. Bradley. 1987. Controlling erosion and sustaining production with no-till systems. *TN Farm Home Sci.* Winter:18-23.

Shelton, C.H., F.D. Tompkins, and D.D. Tyler. 1983. Soil erosion from five soybean tillage systems. *J. Soil Water Cons.* 38:425-428.

Soil Survey Staff. 1975. Soil taxonomy: A basic system of soil classification for making and interpreting soil surveys. *Agr. Handbk. No. 436.* U.S. Dept. Agr., Washington, D.C.

Sojka, R.E., G.W. Langdale, and D.L. Karlen. 1984. Vegetative techniques for reducing water erosion of cropland in the southeastern United States. *Adv. Agron.* 73:155-181.

Trimble, S.W. 1974. *Man-induced soil erosion on the Southern Piedmont, 1700-1970*. Soil Cons. Soc. Am., Ankeny, IA.

Unger, P.W., G.W. Langdale, and R.I. Papendick. 1988. Role of crop residues - improving water conservation and use. pp. 69-100. In: W. L. Hargrove (ed.) *Cropping strategies for efficient use of water and nitrogen*. Spec. Pub. No. 51. Am. Soc. Agron., Madison, WI.

West, L.T., W.P. Miller, G.W. Langdale, R.R. Bruce, J.M. Laflen, and A. W. Thomas. 1991. Cropping system effects on interrill soil loss in the Georgia Piedmont. *Soil Sci. Soc. Am. J.* 55:460-466.

Williams, J.R., R.R. Allmaras, K.G. Renard, L. Lyles, W.C. Moldenhauer, G.W. Langdale, L.D. Meyer, W.J. Rawls, G. Darby, R.R. Daniels, and R. Magleby. 1981. Soil erosion effects on soil productivity: A research perspective. *J. Soil Water Cons.* 36:82-90.

Wischmeier, W.H. and D.D. Smith. 1965. Predicting rainfall erosion losses - a guide to conservation planning. *Handbk. No. 537*. U.S. Dept. Agr., Washington, D.C.

Cover Crops and Rotations

D.W. Reeves

I. Introduction

Recent requirements imposed by government farm policies, shifts in the economics of farming practices, and the public's concern in protecting the environment and conserving our natural resources have created a resurgence of interest in two of the oldest agricultural practices known. These two practices, using cover crops and crop rotations, have been recognized as good management practices since ancient times.

The value of rotations with legumes was recognized by the Chinese over 2,000 years ago (Pieters, 1927). Virgil, in pre-Christian Rome, proclaimed in verse the virtues of fallowing the land from continuous cropping and of rotating small grains with legumes (Gladstones, 1976). Although green manure crops, especially lupin (*Lupinus* spp.), were common in southern Europe long before

1-56670-003-5/94/$0.00+$.50
© 1994 by CRC Press, Inc.

the birth of Christ, crop rotations were unknown in northern Europe until about the 16th century (Pearson, 1967).

Sometime during the 1730's, Lord Townsend of Norfolk County, introduced the Norfolk rotation to England (Pearson, 1967). The Norfolk rotation is a 4-year rotation of wheat (*Triticum aestivum* L.)-turnip (*Brassica rapa* L.)-barley (*Hordeum vulgare* L.)- and red clover (*Trifolium pratense* L.). This rotation was responsible for raising average wheat yields in England from 540 kg ha^{-1} to 1350 kg ha^{-1} by the early 19th century. The Norfolk rotation, or some modification of it, is still in use in northern Europe today.

English settlers carried the knowledge of green manuring to America. A rotation of corn (*Zea mays* L.), wheat, and red clover was described by Thomas Cooper in 1794 as being practiced by the "best" farmers in Pennsylvania (Pieters, 1917). Partridge pea (*Chamaecrista fasciculata* Greene) and cowpea [*Vigna unguiculata* (L.) Walp subsp. *unguiculata*] were grown in rotations in Virginia and Maryland by the end of the 18th century (Pieters, 1927).

Modern farm practices are largely dictated by economics and government policies that affect those economics. As a consequence of these factors, farming has generally become more specialized. Highly capitalized mechanized systems do not lend themselves well to diversified farming practices (Pearson, 1967). As a result of this specialization, cover crops and rotations are not utilized to the extent they once were. Although the current focus on farming practices to "sustain" natural resources and productivity has created a resurgence of interest in these two practices, the interest has not progressed to large scale adoption of the practices. This chapter will review the general principles of crop rotation and cover crops, discuss their advantages and disadvantages in the context of their coordination into current agricultural systems, and outline future research that will facilitate wider adoption of these practices.

II. Principles of Crop Rotation

A systematic or recurrent sequence of crops grown over a number of cropping seasons is a common definition of crop rotation. Cropping season should be considered the unit of time rather than years since in some areas the length of the growing season allows for more than one crop per year. The choice of crops used in rotations is determined by ecology and economics (Pearson, 1967). Ecological limitations are generally not determined by the farmer. They include such things as the crop's suitability to edaphic, biotic and climatic factors. To some degree, however, researchers and farmers have learned to extend the natural ecological niche of crops. Examples of this include growing alfalfa on acid soils modified by liming, irrigating arid regions, chemical control of soil-borne and foliar diseases in warm, humid regions, and northward expansion of the Cotton Belt (*Gossypium hirsutum* L.) through the development and use of "short-season" germplasm and management schemes.

Certain crops, however, have an economic advantage over other crops within a particular area. This economic advantage may be related to environmental adaptation, variations in regional inputs necessary to produce the crop, and government policies and market influences. Pearson (1967) coined the term "comparative advantage" for this edge given by one or more of the above factors to one crop in a particular situation. If the comparative economic advantage of a crop is strong enough, the crop dominates to the extent that it is generally grown in monoculture. If the comparative advantage of one crop is not excessively greater than some others, the farmer is more likely to use crop rotations.

A basic premise to any successful crop rotation is the use of a "hub" crop, i.e., the crop which offers the greatest comparative advantage (Pearson, 1967). The hub crop varies by geographical area and is determined by ecological and economic principles as discussed previously.

The comparative advantage or choice of hub crop is, within limits imposed by climate and soil factors, mainly a function of economic principles. Other principles of crop rotation are based more on agronomic principles than economic principles *per se*. Pearson (1967) summarized these principles to include: i) utilization of both row crops and sod crops; ii) alternation of diverse crop species; iii) fertilizer management based on differential crop nutrient use response and efficiency; iv) a logical sequence and duration of crops; and v) flexibility for risk aversion. From a practical standpoint, crop rotation is best established by dividing the farm into a number of fields of equal size; the number of fields corresponding to the number of growing seasons necessary to complete the rotation cycle (Parker, 1915; Pearson, 1967).

III. Current Rotation Practices in the United States

Ecology and economics not only determine the choice of crops in rotations, but also influence the choice of soil management, i.e., tillage and residue management systems. Allmaras et al. (1991) outlined eleven different tillage management regions (TMR) within the contiguous United States (Figure 1). Two factors, climate and major crops grown, were the factors used to delineate the tillage management regions. Thus, the basic principles that determine the selection of crops grown in a geographical area form not only the basis for grouping regions as to their problems and potential benefits resulting from various residue management systems, but also form the criteria for selection of rotations and cover crops.

Corn (*Zea mays* L.) is the predominant hub crop in the Northern and Southern Corn Belts and Central Great Plains (Allmaras et al., 1991). Approximately 25% of the corn produced in these states is grown continuously, i.e., not rotated with another crop species (Daberkow and Gill, 1989; Gill and Daberkow, 1991). The most common rotation crop for corn is soybean [*Glycine max* (L.) Merr.], comprising about 50% of the acreage planted (Daberkow and

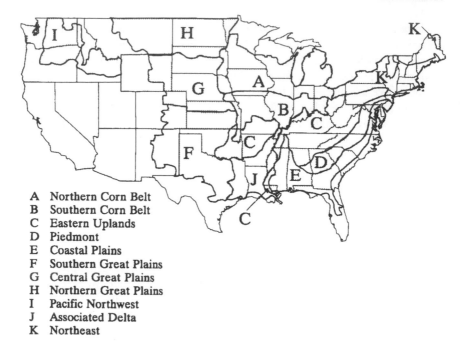

A   Northern Corn Belt
B   Southern Corn Belt
C   Eastern Uplands
D   Piedmont
E   Coastal Plains
F   Southern Great Plains
G   Central Great Plains
H   Northern Great Plains
I   Pacific Northwest
J   Associated Delta
K   Northeast

Figure 1. Eleven tillage management regions in the contiguous United States. (From Allmaras et al., 1991.)

Gill, 1989; Gill and Daberkow, 1991). Alfalfa (*Medicago sativa* L.) and small grain rotations generally comprise the remaining acreage of rotation crops in these areas.

Soybean can also be considered a major hub crop in the Southern Corn Belt tillage management region. In 1990, over 80% of the soybean grown in this area was rotated with corn, 58% in a soybean-corn-soybean rotation; 4% was not rotated, and 3% was grown following a fallow period (Gill and Daberkow, 1991).

In the area where corn and soybean are hub crops, approximately 30% of the corn and 35% of the soybean are planted with conservation-tillage, i.e., where 30% or more residue remains after planting (Taylor and Bull, 1992). The predominant conservation-tillage system in the Southern Corn Belt is mulch-tillage (primary tillage with chisels, disks, or other implements over the entire area of the field before planting to leave at least 30% residue cover)(CTIC, 1991). In certain areas of the Associated Delta, Coastal Plain, Piedmont and Eastern Upland tillage management regions (Figure 1), soybean is also the hub crop; however, in these areas soybean is often rotated in a double-cropping system with wheat. Approximately 30% of the wheat grown in the Southeast in 1990 was double-cropped with soybean (Gill and Daberkow, 1991). Conservation-tillage systems are used on 17% of the land planted to soybean in this area, with no-tillage being the predominant conservation-tillage system used (Taylor and Bull, 1992).

The example of soybean as the hub crop illustrates the interactive relationship between choice of crop rotation, climate, and tillage system utilized. In the northern area of soybean production, corn is the common rotation and mulch tillage is the conservation-tillage system of choice. In the southern area of soybean production, wheat is the rotation crop, and no-tillage is the preferred conservation-tillage technique.

Wheat is a hub crop grown in the Southern and Northern Great Plains tillage management regions (Figure 1) (Allmaras et al., 1991). These areas have mean annual evaporation rates that greatly exceed their annual precipitation (Kilmer, 1982; Allmaras et al., 1991). The role of climate in choosing both the hub crop and rotation is indicated by the fact that approximately 40 to 50% of the wheat grown in this area involves a fallow rotation in order to conserve soil water (Daberkow and Gill, 1989; Gill and Daberkow, 1991). Mulch tillage and reduced tillage (one-pass full-width tillage done at same time as planting/seeding operation) are the conservation-tillage systems used most for wheat in these dry areas (Allmaras et al., 1991; Taylor and Bull, 1992).

Rice (*Oryza sativa* L.) enjoys a comparative advantage on heavy soils in California to the extent that it is grown in continuous monoculture (Daberkow and Gill, 1989). In Arkansas and Louisiana, however, soybean is a rotation crop for rice on about 75% of the planted acreage (Gill and Daberkow, 1991). Because rice is generally planted on flat heavy soils and flooded, conservation-tillage systems are not required to control erosion. Ninety-five percent of the rice in the states in the Associated Delta tillage management region is planted with conventional tillage without the moldboard plow (Taylor and Bull, 1992).

Pest management dictates rotations be used in potato (*Solanum tuberosum* L.) production. In 1990, only 3% of the potatoes grown in the United States were grown following 2 years of potato (Gill and Daberkow, 1991). Small grain preceded potato in about 40% of the acreage planted in 1990. The variety of rotation crops used with potato is greater than for any of the major crops surveyed by Daberkow and Gill (1989) and Gill and Daberkow (1991). This is likely due to the wide distribution of potato production within states in a number of tillage management regions, and to the fact that potato production occurs in areas where livestock and forages are also produced. Traditionally, farms with livestock components most favor integration of crop rotation.

Cotton is the dominant crop in the Southern Great Plains and Associated Delta tillage management regions, and also represents a good portion of the cropland planted in California, Arizona, and the Eastern Upland and Coastal Plain Regions (Figure 1) (Allmaras et al., 1991). The potential profitability of cotton results in a strong comparative advantage, to the point that it is generally grown in continuous monoculture. Daberkow and Gill (1989) cite continuous cotton as the most economic cropping practice in the United States. The role of government farm programs in creating comparative advantage for a crop to the point of discouraging crop rotation is illustrated by the fact that on farms

Table 1. Average wheat yield at Rothamsted, England (1851-1919)

Treatment	Yield (kg ha⁻¹)
Continuous wheat unfertilized	829
Continuous wheat fertilized	1589
Wheat in a Norfolk rotation (turnip, barley, clover, wheat)	1616
Wheat in a Norfolk rotation fertilized	2183

(From Martin et al., 1976.)

enrolled only in the cotton commodity program, over 75% of the cotton is grown continuously (Gill and Daberkow, 1991). Overall, 61% of the cotton grown in 1990 was grown continuously, 6% used corn in rotation, 4% used soybean in rotation, 8% used a fallow period, 2% used a vegetable crop in rotation, and 11% used sorghum (*Sorghum bicolor* L. Moench) in rotation (Gill and Daberkow, 1991).

Not only is cotton the crop grown most continuously, but 97% of cotton is tilled using conventional-tillage systems (Taylor and Bull, 1992). This reliance on conventional-tillage systems is due in part to requirements by some states to destroy cotton plant residues that can serve as food sources for boll weevils (*Anthonomus grandis* Boheman) and bollworms (*Heliothis* spp.). Twenty-two percent of the acreage planted to cotton in 1991 was on land designated as Highly Erodible Land (HEL) (Taylor and Bull, 1992). The combination of reliance on conventional-tillage on 2.36 million acres of HEL (Taylor and Bull, 1992), and lack of crop rotation indicates a critical area for research that needs to be addressed for this major crop.

With the exception of cotton and irrigated corn in Nebraska, continuous cropping is not the prevalent practice for the major crops grown in the United States (Daberkow and Gill, 1989; Gill and Daberkow, 1991). However, the types of rotations are limited, with only 5 to 10 rotations being used on over 80% of the cropland (Daberkow and Gill, 1989).

IV. Advantages of Crop Rotation

Yield increases were the earliest recognized advantage to crop rotations. Long-term wheat yields from the Norfolk rotation at Rothamsted, England illustrate the positive effect of rotation on crop yield (Table 1) (Martin et al., 1976). This "rotation effect" can be attributed to a number of factors, including a reduction in diseases and pests, more efficient weed control, improved water and nutrient use efficiencies, and improved soil physical properties. The underlying effects of crop rotations to improve yield have long been studied; however, it is only recently that research integrating rotations and residue management strategies has sought to better understand the interactive effects of tillage and residue management with rotations. Benefits from rotations have been attributed in part to the following factors:

A. Disease and Pest Control

Crop rotation has long been advised as a means of control of plant diseases (Leighty, 1938). Disease cycles are disrupted when diverse crop species are grown in sequence. For example, in Australia, a single crop of narrow-leaf or blue lupin (*Lupinus angustifolius* L.) reduced the infection rate of common root rot (*Fusarium* and *Bipolaris* spp.) in wheat compared to wheat in monoculture by approximately half (Reeves, 1984). Likewise, wheat in rotation with lupin reduced *Fusarium* spp. and leaf brown spot [*Pleiochaeta setosa* (Kirchn.) Hughes] infection in lupin (Reeves, 1984). Rotations are an effective method for control of a number of wheat diseases in the Pacific Northwest tillage management region (Cook, 1986).

Nematode control through crop rotation is especially effective. As more chemicals for control of nematodes are removed from the market due to environmental concerns, reliance on crop rotation for nematode control will increase. Kinloch (1983) found that yields of soybean varieties both resistant and susceptible to southern root-knot nematode [*Meloidogyne incognita* (Kofoid and White) Chitwood] were increased when grown in rotation with corn; however, only yields of susceptible varieties declined with reduced length in rotation. Inclusion of bahiagrass (*Paspalum notatum* Flügge) sod in rotation with peanut (*Arachis hypogaea* L.) is an effective control for root-knot [*Meloidogyne arenaria* (Neal) Chitwood] and lance (*Hoplolaimus coronatus* Cobb) nematode (Norden et al., 1977; Rodríguez-Kábana et al.,1988; Dickson and Hewlett, 1989). A 2-year rotation of bahiagrass reduced populations of *M. arenaria* and soybean cyst nematode (*Heterodera glycines* Ichinohe) to undetectable levels in Alabama and increased yield of soybean up to 114% (Rodríquez-Kábana et al., 1991).

Crop rotation is effective in controlling insect pests with narrow host compatibility and short migration distances (Francis and Clegg, 1990). Insects like corn rootworm (*Diabrotica* sp.), corn root aphid (*Anuraphis maidiradicis* Forbes), billbugs (*Sphenophorus* spp.) and wireworms (*Agriotes mancus* Say, *Horistonotus uhlerii* Horn, *Melanotus* spp., and *Aeolus mellillus*) can be controlled using crop rotations (Dicke and Guthrie, 1988; Luna and House, 1990).

Although the benefits of rotation *per se* are well represented in the literature, the interactive effects of rotations and tillage systems is not as well researched. In Alabama, conservation tillage systems (no-tillage and strip-tillage) in combination with a rotation with corn resulted in a consistent soybean yield increase compared to conventional tillage and continuous cropping (Edwards et al., 1988). The increase was due to a build up of soybean cyst nematode under conventional tillage and continuous soybean. In Indiana, sudden death syndrome of soybean (SDS) as a result of *Diaporthe* and *Phialophora* spp. occurred at the rate of 11% in continuous soybean, 6% in a corn-soybean rotation, and 2% in a wheat-corn-soybean rotation (von Qualen et al., 1989). Yield reductions were commensurate with SDS occurrence. In continuous soybean, SDS occurred more frequently with no-tillage than with conventional or chisel tillage. In contrast to these findings, other reports show reductions of some disease organisms with conservation-tillage (Rothrock, 1987; Herman, 1990). In Herman's study, infestation of wheat with take-all [*Gaeumannomyces graminis* (Sacc.) von Arx & Olivier var. *tritici* (Walker)] was greatest following alfalfa under conventional-tillage compared to minimal tillage. However, Herman concluded that tillage as a means of control for take-all was less effective in a crop rotation than in monoculture.

B. Weed Management

Leighty (1938) summarized the impact of crop rotation on weed control when he said "No other method of weed control, mechanical, chemical, or biological, is so economical or so easily practiced as a well-arranged sequence of tillage and cropping." Given an expanded definition of the words "sequence of tillage" as was known in 1938 to include practices involved in residue management known today, these words still ring true. Specific weeds are generally associated with specific crops (Froud-Williams, 1988). Rotation of crops breaks the life cycles of weeds adapted to the narrow ecological niche imposed by continuous cropping. Selective pressures on weeds, including crop competition, pathogens and pests, herbicide tolerance, fertility factors, and tillage, are reduced when rotation is not practiced. Rotation prevents dominance of a relatively narrow range of difficult to control weed species. Rotations are especially effective in controlling crop mimics like shattercane (wild x cultivated *Sorghum bicolor* crosses) in corn and grain sorghum (Francis and Clegg, 1990) and weed beet [*Beta vulgaris* L. ssp. *maritima* (L.) Arc./Thell] in sugarbeet (*Beta vulgaris* L.) (Froud-Williams, 1988).

Reduced tillage and consequent reliance on herbicides for weed control results in reduced weed species diversity but denser populations of highly specialized species, especially annual and perennial grasses (Peterson and Russ, 1982; Froud-Williams, 1988). Thus, the importance of crop rotation as a means of weed control is even greater in reduced tillage systems than in conventional systems.

C. Erosion Control

The development and wide scale use of herbicides and inorganic sources of N during the 1940's moved agriculture away from sod-based or meadow rotations to continuous cropping of row crops. These sod-based or hay-crop rotations are very effective in reducing soil erosion (Leighty, 1938; Mannering et al., 1968; Carreker et al., 1977; Langdale et al., 1991). Obviously, a key factor in residue management strategies is the type of residue left by a preceding crop. Previous crop residues are a major component of the cover and management factor (C) of the Universal Soil Loss Equation (USLE) (Wischmeier and Smith, 1978). As an example, the C factor for corn mulch, with 40% soil coverage and no-tillage is 0.21 while for soybean mulch under the same conditions it is 0.26 (Wischmeier and Smith, 1978). van Doren, Jr. et al. (1984) found that erosivity of corn-soybean rotation averaged over 18 years was 45% greater than with continuous corn. At equal residue cover, however, no-tillage was as effective in reducing erosion following soybean as following corn. On the Texas High Plains, reduced tillage operations resulted in greater yields of cotton and sorghum (Lacewell et al., 1989). Monoculture cotton and sorghum cropping systems produced the greatest wind erosion and 2- or 3-year rotations with wheat reduced average wind erosion to less than 13.5 Mg ha^{-1} per year.

The interactions of crop residue quantity, quality (C:N ratio), crop sequence, and tillage practices are complex, but conservation-tillage with intensive cropping sequences can dramatically reduce erosion losses (Mills et al., 1986; Langdale and Wilson, 1987). The most intensive cropping systems, i.e., those that maintain a crop year round, involve winter cover crops, either small grain cash crops or winter annual legumes.

D. Soil Physical Conditions

Yield increases arising from crop rotation are often a direct result of increased soil productivity from improvement of soil physical properties. The overall effect of rotation on soil physical properties is influenced by the interactive nature of the quantity and quality of residue produced both above- and below-ground and the management of these residues.

Crop species vary considerably in their ability to modify soil physical and chemical properties. Plant roots exert strong influences on soil conditions and crop species vary considerably in their rooting distribution (Table 2). Uhland (1949) reported that deep-rooted crops such as kudzu [*Pueraria lobata* (Willd.) Ohwi] and alfalfa increased the infiltration rate to a depth of 45 cm compared to cropping with cotton. More recently, Meek et al. (1990) reported that cotton planted with no-tillage and minimum tillage maintained macropores produced by alfalfa, producing high infiltration rates. In Australia, a wheat yield response to rotation with narrow-leaf or blue lupin of 100 kg ha^{-1} was attributed to lupin roots acting as "biological plows" in a compacted soil (Henderson, 1989). Elkins et al. (1977) reported that cotton yields doubled following bahiagrass due to an eight-fold increase in pores greater than 1.0 mm diameter within a compacted soil layer. These examples indicate that rotations have the ability, in some instances, to eliminate tillage operations.

Table 2. Rooting depth and lateral spread of roots for some important agronomic crops

Plant	Maximum rooting depth	Effective rooting depth	Lateral spread
	----------------cm----------------		
Oat (*Avena sativa* L.)	200	150	25
Sugar beet	180	120	45
Turnip	168	150	75
Bermudagrass (*Cynodon dactylon* L.)	245	200	--
Soybean	225	200	50
Barley	195	135	30
Alfalfa	610	300	15
Bahiagrass	245	200	--
Garden pea	90	90	60
Rye	230	150	25
Potato	150	90	40
Sorghum	180	180	60
Wheat	200	150	15
Corn	188	180	100

(From Hanson, 1990.)

Roots affect soil structure by their influence on aggregation. Roots and associated fungal hypae bind large aggregates and root exudates stabilize microaggregates (Monroe and Kladivko, 1987; Habib et al., 1990; Dexter, 1991). Differences in aggregate formation and stability have been reported for different crops. The often reported increase in aggregation following sod crops has been attributed to the high density of roots and consequent water extraction cycles under these crops (Dexter, 1991). In some instances, as with corn, roots have been reported to reduce aggregate stability (Reid and Goss, 1982; Reid et al., 1982). In Kansas, sorghum cropping produced smaller, less dry-stable but more wet-stable aggregates than cropping with wheat (Skidmore et al., 1986). In that study, differences were also noted in other physical properties. Saturated hydraulic conductivity was from five to ten times greater following sorghum than wheat. Although sorghum cropping improved soil tilth characteristics, it also resulted in a high wind erodibility index compared to wheat. Bulk density and soil strength are also affected by cropping sequence and residue management (Carreker et al., 1968; Bruce et al., 1990; McFarland et al., 1990).

Maintenance of soil organic matter is the key to improved soil physical properties. Residue management strategies that increase cropping intensity and reduce incorporation of residues have the greatest impact on soil organic matter (Havlin et al., 1990; Langdale et al., 1990; Wood et al., 1990; Wood et al., 1991; Edwards et al., 1992). Increased soil C is associated with better aggregation, infiltration, and other soil properties that result in a more productive soil.

The literature is extensive regarding the effect of tillage and residue management on soil physical properties. In addition, many reports have dealt with the effect of cover crop-, sod- or meadow-based rotations on soil C and resultant changes in soil physical properties. However, the interactive effects of cropping sequence and tillage/residue management have not been extensively studied. Crookston and Kurle (1989) conducted an experiment to determine the role of crop residues *per se* in providing the "rotation effect". They used a 3-yr sequence of corn and soybean cropping in combination with different residue removal/transfer schemes following grain harvest of both crops to determine if the "rotation effect" was due to the decomposition of above-ground residues. They found that the positive influence of corn preceding soybean and vice versa was not due to the above-ground residues produced by either crop. Bruce et al. (1990) looked at the interaction of crop rotation and tillage on soil physical properties on a Piedmont soil in Georgia. They measured greater sorptivity, aggregate stability, higher air-filled pore space, and lower bulk density after two or more years of sorghum than after soybean. Tillage negated or masked the crop rotation effect on soil physical properties. Grain yield responses of soybean and sorghum were commensurate with changes in soil physical conditions brought about by tillage and crop rotation. On a Ships clay soil (Udic Chromustert) in Texas, reduced soil bulk density in the surface 76-mm following a sorghum-wheat-soybean rotation as compared to continuous soybean or a wheat-soybean rotation was attributed to greater residue production with that cropping sequence. Tillage system, however, had no effect on bulk density (McFarland et al., 1990). At the 100- to 200-mm depth, soil strength was greater in the sorghum-wheat-soybean rotation under conventional tillage compared to no-tillage due to the increased equipment traffic in this intensive cropping system with conventional tillage.

More research needs to be conducted regarding crop rotation-tillage/residue management interactions. A more comprehensive understanding of soil and crop specific responses to crop rotation and tillage/residue management practices is critical to improving economies of production. Griffith et al. (1988), for example, reported that on a low organic matter, poorly drained silt loam soil in Indiana (Typic Ochraqaulf), no-tillage compared to moldboard plowing resulted in equivalent yield potential for continuous corn, but increased yield potential when corn was rotated with soybean. The increased yield potential with rotation was linked to improved soil physical conditions. Conversely, on a dark poorly drained silty clay loam high in organic matter (Typic Haplaquoll), continuous corn yields with no-tillage were reduced an average of 9.2% compared to moldboard plowing due to reductions in soil temperature. However, corn yields following soybean with no-tillage were only reduced 2.6% compared to moldboard plowing. On this soil, ridge-tillage and crop rotation produced equivalent yields to corn following soybean that was moldboard plowed. Although more research is needed, the examples discussed here suggest that crop rotation greatly influences the effect of tillage on soil physical conditions and crop response. Tillage can mask crop rotation responses and rotation can alleviate potential adverse effects of reduced tillage on certain soils.

E. Efficient Use of Soil Water and Nutrients

The use of legumes and differences in crop rooting patterns are largely the basis for two principles of crop rotation listed by Pearson (1967), i.e., fertilizer management based on differential crop nutrient use response and efficiency, and the use of a logical sequence and duration of crops. Obviously, the use of a legume in a rotation can affect N fertilizer requirements. The fixation of atmospheric N by legumes is the cornerstone of meadow-based rotations used since before the birth of Christ. The fertilizer-N requirement for a crop following a legume is not necessarily reduced, however, because the yield potential of a crop grown in rotation is often increased, resulting in a response to higher N rates. Crop rotation can increase yield potential and consequent N fertilizer efficiency (Peterson and Varvel, 1989a; Karlen et al., 1991). Nitrogen fertilization minimized differences between continuous vs. rotated (with soybean) grain sorghum yields in Nebraska (Peterson and Varvel, 1989b), and soybean and sorghum grown in a 2-yr rotation maintained high yields without additions of N fertilizer. For an in-depth review of N use efficiency and crop rotation the reader is referred to Pierce and Rice (1988).

Soil nutrients other than N are affected by crop rotation and residue management. Reduced soil disturbance and inversion generally results in a stratification of nutrients (Eckert, 1985; Touchton and Sims, 1987; Dalal et al., 1991; Robbins and Voss, 1991). Root morphology and distribution, as well as rhizophere activity, affect uptake of nutrients. Roder et al. (1989) reported that rooting of both soybean and grain sorghum was reduced following a previous crop of soybean vs. sorghum. Johnson et al. (1992) linked the spore population of mycorrhizal fungi associated with corn or with soybean to reductions in yield and tissue concentrations of P, Cu, and Zn in both crops when either crop followed itself. With no-tillage, crop rotation affected the distribution P, K, Mg, and Ca within the 20-cm depth of an Alfisol in Nigeria (Lal, 1976). Fibrous rooted grass species are generally more effective than tap-rooted species in extracting P (Mays et al., 1980). Conversely, some tap-rooted crops, like white (*Lupinus albus* L.) and blue lupin, can secrete large amounts of organic acids that result in increased availability of P (Gardner and Parbery, 1982; Tadano and Sakai, 1991). This increased P availability can be carried over to succeeding crops grown in rotation (Meredith, 1992). Deep rooting and K extraction by blue lupin improved the K status in the surface 10-cm of a deep sandy soil in Australia, with a subsequent yield response by a following wheat crop (Rowland et al., 1986). The K fertilizer required by a cropping system is dependent on the crops grown in a rotation and the sequence of crops in the rotation determines the time that K should be applied in order to maximize the efficiency of the application (Pretty and Stangel, 1985).

As discussed earlier, tillage/residue management and crop rotation interact with varied influences on soil physical properties, most importantly those that relate to soil water. In semiarid regions the effect of rotation and crop residue management strategies on soil water storage and extraction are critical. In the Southern Great Plains tillage management region, decreasing tillage intensity and increasing surface residues of wheat increased soil water storage which increased sorghum yields (Unger, 1984). Including sunflower (*Helianthus annuus* L.) in the rotation allowed extraction of water deeper within the profile, increasing the total amount of water available for crop production in a fallow-wheat-sorghum-sunflower-wheat rotation. In general, increased soil C, whether from increased surface residues as a result of tillage reductions or from rotations with grasses and legumes, results in greater soil aggregation and infiltration. The end result of these processes is greater soil water storage. Also, the most practical way to increase water use efficiency (WUE) is to increase yield (Pendleton, 1966). Since crop rotations generally increase yield potential, they thus then result in increased WUE.

V. Cover Crops

Pieters (1927) credits Richard Parkinson (1799) as the first person to stress the idea of cover crops. Pieters quotes Parkinson as saying, "... it is seen how earnest my wish is that the surface of the ground should at all times, winter and summer, be well covered, whenever it possibly can be accomplished." Cover crops are defined as crops grown specifically for covering the ground to protect the soil from erosion and loss of plant nutrients through leaching and runoff (Parker, 1915; Pieters and McKee, 1938). Long-term rotations of grass or legume sods have been treated as cover crops in discussions by some workers; however, for the purposes of this chapter, only crops grown in the off-season with an annual planting of a cash crop will be defined as cover crops.

Cover crops are in essence short-term rotations. Traditionally, cover crops were turned under and incorporated before planting of the cash crop; however, the increased emphasis on residue management as a means for reducing soil erosion has led to greater use of cover crops in conservation-tillage systems. Also, in tillage management regions with more than 180 continuous frost-free days, double-cropping a summer crop behind a small grain winter crop is possible (Allmaras et al., 1991). The small grain winter crop thus serves dual purposes, i.e., as a cash crop and as a cover crop.

In residue management systems, a cover crop must meet certain requirements: i) it should be easy to establish; ii) it should have a rapid growth rate so as to provide ground coverage quickly; iii) it should produce a sufficient quantity of dry matter for maintenance of residues; iv) it should be disease resistant and not act as a host for diseases of the cash crop; v) it should be easy to kill; vi) it must be economically viable. The degree that a specific cover crop meets these specifications is dependent on the soil, climate, and succeeding cash crop, as well as characteristics of the cover crop itself.

During the early part of the twentieth century in the United States, in addition to winter cover crops, summer annuals like cowpea, soybean, and velvetbean [*Mucuna deeringiana* (Bort) Merr.] were grown as green manures to be turned under at the end of the season for soil improvement (Pieters, 1927). Currently, only winter-season cover crops are currently used in temperate and subtropical zone cropping systems. Both legumes and grasses (small grains) are used as winter cover crops. Many of the advantages and disadvantages of cover crops are common to both small grains and legumes. Where differences in effects from the two types of covers occur, they are largely related to fixation of N by legumes and resultant differences in N content of residues between the two groups of cover crops.

A. Small Grain Cover Crops

For small grain covers, the residue N content depends on soil N availability, which is dependent on the amount of residual soil N as well as the mineralization rate (Wagger and Mengel, 1988). Residual N is mainly dependent on previous crop N utilization. In addition, the overall growth rate and stage of phenological development that the cover is terminated greatly influences the N uptake of small grain cover crops. Nitrogen content of small grain cover crop residues varies greatly, but generally ranges from 25 to 50 kg N ha^{-1} (Table 3). Often overlooked in reports of N uptake by cover crops is the amount of N present in root residues. For small grains, reports of N present in roots range from 8 to 42% of total N uptake by the cover (Mitchell and Teel, 1977; Scott et al., 1987; Reeves et al., 1993). The C:N ratio of small grain residue is mostly dependent on total dry matter produced and time of termination. Early termination of the cover results in a narrower C:N ratio in the residue, but the total residue produced is reduced. If killed too early, the narrower C:N ratio results in rapid decomposition of the residue, reducing ground coverage. In practice, however, small grain cover crops are usually killed at a stage of development that results in a wide C:N ratio, usually exceeding 30:1 (Table 3). This wide C:N ratio results in an initial, if not persistent, immobilization of N during the cropping season (Aulakh et al., 1991; Doran and Smith, 1991; Somda et al., 1991; Torbert and Reeves, 1991).

Initially, the potential for denitrification and immobilization of N is frequently greater with no-tillage than for conventional-tillage (Doran 1980a; Doran 1980b; Rice and Smith, 1984). This coupled with the wide C:N ratio of small grain cover crop residues dictates the importance of proper N management in these type systems. The use of starter fertilizer to supply 25 to 30 kg N ha^{-1} to crops planted with conservation-tillage behind small grain cover crops has been shown to be a good management practice (Reeves et al., 1986; Touchton et al., 1986;

Table 3. Nitrogen content of small grain cover crops

Cover crop	N content (average)	C:N	Reference
	kg ha[-1]		
Wheat	--	82	Aulakh et al., 1991
	--	97	Somda et al., 1991
	74	--	Decker et al., 1987
	32	22[b]	McVay et al., 1989
	21[a]	18[b]	Scott et al., 1987
Barley	69	--	Decker et al., 1987
Rye	100	25[b]	Hoyt, 1987
	42[a]	29[b]	Reeves et al., 1993
	42[a]	26[b]	Scott et al., 1987
	85	35	Wagger, 1989
	60[a]	40	Mitchell and Teel, 1977
	24	--	Brown et al., 1985
	36	38[b]	Ebelhar et al., 1984
	14	57[b]	Blevins et al., 1990
	38	42[b]	Hargrove, 1986
	13	54	Huntington et al., 1985

--, Data not available; [a]includes roots; [b]calculated at C=40% dry matter.

Hairston et al., 1987; Reeves et al., 1990; Howard and Mullen, 1991). Although yield increases from starter N applications are dependent on rainfall, accompanying tillage to disrupt compacted soil layers, and crop, they occur frequently enough to justify the practice. In addition, early-season growth of the cash crop is almost always enhanced with starter N applications, providing more rapid canopy coverage of row middles, lessening weed competition.

B. Legume Cover Crops

The N content of legume cover crop residues varies with species, residual soil N, adaptability to specific soil and climatic conditions, and time of termination of growth. Nitrogen contents of a number of winter legume cover crops are listed in Table 4. The values reported in Table 4 are averaged across sites, years and management practices for each reference. In these cases, the N content ranged from 36 to 226 kg ha[-1], with the average N content of above-ground residues being 120 kg ha[-1]. The C:N ratio varied from 25:1 to 9:1, but in all but two reports, the ratio is well below 20:1, the guideline threshold where rapid mineralization of the N in the residue would occur.

In most studies, the N content of root residues has not been determined. In Virginia experiments, roots of an Austrian winter pea (*Pisum sativum* L.) cover

Table 4. Nitrogen content of legume cover crops (values averaged for studies over years, locations, and management practices)

Cover crop	N content	C:N[a]	Reference
	(kg ha⁻¹)		
Arrowleaf clover	131	15	Fleming et al., 1981
(*Trifolium vesiculosum* L.)			
Austrian winter pea	161	11	Hoyt, 1987
	68	9	Neely et al., 1987
Berseem clover	67	16	McVay et al., 1989
(*Trifolium Alexandrinum* L.)			
Bigflower vetch	67	13	Blevins et al., 1990
(*Vicia grandiflora* Scop)	60	13	Blevins et al., 1990
	134	10	Hargrove, 1986
Common vetch	85	23	Touchton et al., 1984
(*Vicia sativa* L.)	153[b]	16	Reeves et al., 1993
Crimson clover	170	17	Hargrove, 1986
	108	13	McVay et al., 1989
	88	11	Brown et al., 1985
	126	15	Wagger, 1989
	56	17	Ebelhar et al., 1984
	114	11	Neely et al., 1987
	129	15	Hoyt, 1987
	94	25	Touchton et al. 1984
	163	16	Fleming et al., 1981
	103	13	Blevins et al., 1990
Hairy vetch	104	15	Brown et al., 1985
	209	10	Ebelhar et al., 1984
	158	9	Hoyt, 1987
	125	13[c]	Huntington et al., 1985
	128	11	McVay et al., 1989
	127	9	Neely et al., 1987
	36	8	Power et al., 1991
	161	10	Wagger, 1989
	114	14	Hargrove, 1986
Subterranean clover	226	19	Reeves and Mask, 1992
White lupin			

[a]Calculated as C=40% dry matter; [b]includes roots; [c]actual value reported.

crop contained 47 kg N ha⁻¹, crimson clover (*Trifolium incarnatum* L.) roots 77 kg N ha⁻¹, vetch (species not reported) roots 118 kg N ha⁻¹, and ryegrass (*Lolium multiflorum* Lam.) roots 138 kg N ha⁻¹ (McVickar et al., 1946). The soil core technique used to measure roots in this study probably over-estimated the N content of the roots. It is unlikely that ryegrass roots, for example, contained

79% of the total amount of N in the plant as reported. Reeves et al. (1993) reported that 16% of the N in a crimson clover cover crop was found in roots. Mitchell and Teel (1977) reported that 9 to 13% of the N in small grain-legume cover crop mixes was found in roots. The fraction of N in nonharvested roots and crowns of alfalfa, red clover and birdsfoot trefoil (*Lotus corniculatus* L.) forages preceding a corn crop was reported to be 31, 22, and 26%, respectively, of the total plant N (Sheaffer et al., 1991). Kirchmann (1988) reported considerable species variation in the total amount of N partitioned to roots of red clover, white clover (*T. repens* L.), Persian clover (*T. resupinatum* L.), black medic (*Medicago lupulina* L.), Egyptian clover (*T. alexandrium* L.) and subterranean clover (*T. subterranean* L.). The fraction of the total amount of N found in the roots varied from 3% in Persian clover to 45% in white clover. The contribution of N in roots of legume cover crops is not insignificant, and more research needs to be conducted as to the role N from roots plays in residue management strategies.

Management factors strongly influence the N content of legume cover crops and the contribution of N available to the following cash crop. Practices that promote early establishment, i.e., early planting, interseeding, or natural reseeding, result in greater dry matter and consequent N production (Brown et al., 1985; Oyer and Touchton, 1990). The time of termination of the cover crop also affects the N content of the residues. Wagger (1989) reported that delaying the kill date of rye (*Secale cereale* L.) 2 weeks beyond anthesis, crimson clover 2 weeks beyond 50% bloom, and hairy vetch (*Vicia villosa* Roth) 2 weeks beyond 25% bloom increased cover crop dry matter of rye by 39%, clover 41%, and vetch 61%. Corresponding increases in N content were 14% for rye, 23% for clover, and 41% for vetch.

Of primary importance is the use of an adapted species to the tillage management region. The use of winter annual legumes cover crops is limited by extremes in minimum temperature in the northern United States and by their competition with the cash crop for soil water in the western United States. Limited research has shown there is potential for use of winter annual legume cover crops in rotations in semiarid and northern temperate environments (Auld et al., 1982; Badaruddin and Meyer, 1989; Gilley et al., 1989; Power, 1991; Power et al., 1991). In these areas, however, legumes are utilized in the higher rainfall areas and are mainly used as substitution for fallow in rotations. Tillage management regions with the greatest potential for using winter annual legume cover crops are the Coastal Plain, Piedmont, Associated Delta, Eastern Uplands, and Southern Corn Belt (Figure 1).

Crimson clover and hairy vetch are time-proven legume cover crops. They represent the standards by which other species are compared in most research. Crimson clover is earlier to produce maximum dry matter and seed than hairy vetch but it is not as cold hardy. Dry matter production is the major determining factor in N production by winter legume cover crops (Holderbaum et al., 1990). The transition zone where performance of hairy vetch surpasses that of crimson clover corresponds to the plant hardiness zone where the average annual

Figure 2. Transition zone for the adaptability of hairy vetch and crimson clover cover crops. The average annual minimum temperature in the zone is −15 to 17.7°C. Hairy vetch is more reliable than crimson clover north of this zone. (Adapted from USDA Plant Hardiness Map, 1990.)

minimum temperature is -15 to -17.7°C (5 to 0°F) (Figure 2). North of this zone, hairy vetch performs better while south of this zone the earlier growth of crimson clover gives it an edge over hairy vetch. Soil factors also play a role in adaptability of legume cover crops. Crimson clover performed better in Tennessee on poorly drained soils and on well limed soils than hairy vetch, but hairy vetch and subterranean clover were more tolerant of soil acidity (Duck and Tyler, 1987).

C. Nitrogen Fertilizer Equivalence

The N contribution from small grain cover crops generally is negative (Brown et al., 1985; Wagger, 1989; Reeves and Touchton, 1991a; Torbert and Reeves, 1991) due to the wide C:N ratio of the residue. Nitrogen fertilizer equivalence for legume cover crops has been extensively reviewed (Hoyt and Hargrove, 1986; Smith et al., 1987; Frye et al., 1988). Nitrogen fertilizer equivalences reported in these reviews ranged from 15 to 200 kg ha^{-1}, with the typical values being about 60 to 100 kg ha^{-1}.

The N contribution is dependent on environmental and management factors that affect the amount of dry matter produced by the cover crop as well as the environmental and management factors that influence the yield potential of the cash crop. Conditions that increase the yield potential of the cash crop will increase the response to N. The N contribution from the legume to the cash crop is generally calculated as the N fertilizer needed following a small grain cover crop (or no cover crop) to obtain a yield equivalent to that following the legume cover crop without N fertilizer. Although practical, this method of calculating N contribution from the cover crop does not discriminate between the N contribution effect of the cover crop and other effects of the cover crop (Smith et al., 1987; Frye et al., 1988). As will be discussed later, cover crop effects that influence yield of following crops include changes in rooting, soil structure, soil water, soil temperature, weed control, etc. Smith et al. (1987) suggests that N accumulation differences rather than yield would be a better indicator of N contribution of the cover crop. Hargrove (1986) suggested that grain N content of the following crop is a more appropriate measure of the N contribution of the cover crop. Russelle et al. (1987) proposed a method of discriminating between N contribution and other, i.e., rotation, effects by using yield-N uptake response curves of a crop grown continuously and N uptake of the crop with and without rotation with a legume. Although the method requires some assumptions, it is practical and does not require the use of N isotopes to determine N contribution of a legume to a following crop.

Only a few studies with ^{15}N have been used to determine the contribution of legume crop residues to following crops. In pot and field studies, recovery of labeled legume residues has ranged from 5 to 28% (Norman and Werkman, 1943; Ladd et al., 1983; Azam et al., 1985; Westcott and Mikkelsen, 1987). There is a real scarcity of isotopic research dealing with the effect of tillage on availability of legume cover crop N. The primary work in this area was done in Kentucky with ^{15}N labeled hairy vetch (Varco et al., 1989). Recovery of vetch ^{15}N by a succeeding corn crop averaged 32% with conventional tillage and 20% with no-tillage. Residual ^{15}N recovery the second year after labeled legume residue was applied to plots was 7% with no-tillage and 3% with conventional tillage.

Incorporation of legume cover crop residues results in more rapid mineralization than when residues are left on the soil surface (Wilson and Hargrove, 1986; Groffman et al., 1987); consequently, legume-residue N may not be available to the succeeding grain crop during the early part of the growing season. In Kentucky, the majority of N mineralized from decomposing residues of hairy vetch left on the soil surface did not become available to corn grown with no-tillage until after silking (Huntington et al., 1985). In another study, N uptake prior to silking was greater with conventional tillage; however, the potential for greater uptake of N from hairy vetch residues during grain fill of corn was found to be greater, dependent on rainfall, with no-tillage than with conventional tillage (Varco et al., 1989). Due to the initial lag in availability of N from legume cover crop residues, fertilizer N applications should be applied at planting in conservation-tillage winter annual legume systems (Reeves et al., 1993). Splitting N applications to corn grown in these systems, as is generally recommended for conventional-tilled corn grown without legume cover crops, is not necessary (Reeves et al., 1993).

D. Benefits of Cover Crops

1. Improved Soil Physical Conditions

Strictly speaking, cover crops are grown to protect the soil from erosion and loss of plant nutrients while green manures are crops that are grown with the purpose of improving soil productivity. Green manures have traditionally been incorporated. In reality, however, the use of cover crops in residue management strategies offers benefits often attributed to green manures. Cover crops affect soil physical properties primarily due to the production of biomass which serves as the source of soil organic matter and substrate for soil biological activity (Bruce et al., 1991) . As discussed previously, certain crops can also physically modify the soil profile (Uhland, 1949; Elkins et al., 1977; Kemper and Derpsch, 1981; Wilson et al., 1982; Henderson, 1989; Meek et al., 1990), as well as affect soil structure through their influence on soil aggregation (McVay et al., 1989; Dexter, 1991).

It is well established that cover crops can maintain or increase soil C and N (Pieters and McKee, 1938; Lewis and Hunter, 1940; Wilson et al., 1982; Hargrove, 1986; Utomo et al., 1987; McVay et al., 1989; Keisling et al., 1990). The benefits of cover crops in maintaining or increasing soil organic matter are negated with tillage (Utomo et al., 1987). Increased soil C is largely responsible for the changes in physical properties associated with cover crops. The most important agronomic factor affected is associated with soil water relationships. Cover crops frequently result in greater infiltration of water, due to direct effects of the residue coverage or to changes in aggregation and formation of macropores by roots.

In Alabama, after 5 years of stripped-tilled corn, a reseeding crimson clover cover crop increased the percentage of water stable aggregates over winter fallow from 44% to 55% on an Appalachian Plateau soil (Typic Hapludult), and from 40% to 49% on a Coastal Plain soil (Typic Kandiudult) (D. W. Reeves, unpublished data). Similar results were reported for a Piedmont soil in Georgia (Bruce et al., 1992). After 5 years, average water stability of aggregates in the 0- to 15-cm depth with no-tillage grain sorghum planted into crimson clover residue was 53% greater than with conventional-tillage grain sorghum following winter fallow and 44% greater than with conventional-tillage soybean after winter fallow. A previous report from this site showed that grain sorghum grown with no-tillage in combination with a crimson clover cover crop resulted in 89% water-stable aggregates while conventional tillage without a cover crop resulted in 58% water-stable aggregates (Bruce et al., 1991). On a Limestone Valley soil (Typic Hapludult) in Georgia, cover crops had no effect on percentage water-stable aggregates, but on a Coastal Plain soil (Rhodic Paleudult) cover crops increased the percentage of aggregates (McVay et al., 1989). Legume cover crops (hairy vetch and crimson clover) tended to increase aggregate stability more than a wheat cover crop. Increased soil organic matter and aggregation with cover crops as opposed to fallow is the result of increased biomass production. This increased biomass production is generated by the cover crop itself, as well as the indirect effect of the cover crop in increasing biomass yield of the following cash crop.

Increases in soil porosity due to the action of roots, especially under reduced tillage conditions, are frequently cited as the cause of greater infiltration. In addition, cover crop residues left on the soil surface reduce surface sealing of soils, increasing infiltration. In Brazil, a range of cover crops including winter annual legumes and rape *(Brassicas* spp.) increased infiltration rates on Oxisols up to 416% and up to 629% on Alfisols compared to wheat stubble (Kemper and Derpsch, 1981). The infiltration increase was attributed to biological loosening by the cover crop root system. In the Georgia Piedmont, infiltration rate, averaged across a slightly, moderately, and severely eroded site, measured after 4 or 5 yr of no-till grain sorghum planted into a crimson clover cover crop was 100% greater than planting either grain sorghum or soybean with conventional tillage following winter fallow (Bruce et al., 1992). Wilson et al. (1982) reported that a number of cover crops increased infiltration and macroporosity on an eroded Alfisol in Nigeria. Legume covers were especially effective. On a Coastal Plain soil in Georgia, infiltration in no-tillage grain sorghum (measured using a sprinkling infiltrometer) averaged 58.4 mm h^{-1} following a hairy vetch cover crop, 42.3 mm h^{-1} following wheat cover, and 37.8 mm h^{-1} following winter fallow (McVay et al., 1989). The measurements were made after 3 years of cropping. Hairy vetch also increased infiltration compared to winter fallow in a no-tillage corn system on a Limestone Valley soil. Similar results showing the benefit of winter legume cover crops with no-tillage cotton are seen in Figure 3 (Touchton et al., 1984). In contrast, on an Aquic Paleudult in North Carolina, Wagger and Denton (1989) found no differences in soil porosity and saturated hydraulic conductivity (K_{sat}) due to cover crops of wheat and hairy vetch compared to winter fallow in a strip-tillage corn experiment. In untrafficked interrows, however, there was a nonsignificant trend for higher K_{sat} with cover crops, especially wheat. Coefficients of variation were high in this experiment, making it difficult to separate treatment differences among cover crops. In Alabama, a crimson clover cover crop increased K_{sat} of undisturbed soil cores (6-cm depth) from a 5-year experiment with strip-tilled corn 342% on a Typic Hapludult and 214% on a Typic Kandiudult compared to winter fallow (D.W. Reeves, unpublished data). In Arkansas, Keisling et al. (1990) measured K_{sat} of a Typic Hapludalf-Aeric Ochraqualf soil association that had been cropped to cotton for 17 years with and without winter cover crops. A rye-hairy vetch cover crop increased K_{sat} 166% in the 0 to 5 cm-depth, 194% in the 5 to 10 cm-depth, and 359% in the 10 to 15 cm-depth compared to no cover crop. The increase in K_{sat} with depth in this experiment was likely due to incorporation of cover crop residue, as the cotton was conventionally-tilled.

Cover crops also affect soil bulk density and soil strength. Since roots occupy the soil space previously occupied by soil pores, they must displace soil particles, increasing the bulk density of soil adjacent to the root (Glinski and Lipiec, 1990). However, as roots die and decompose they leave macropores in

146

D.W. Reeves

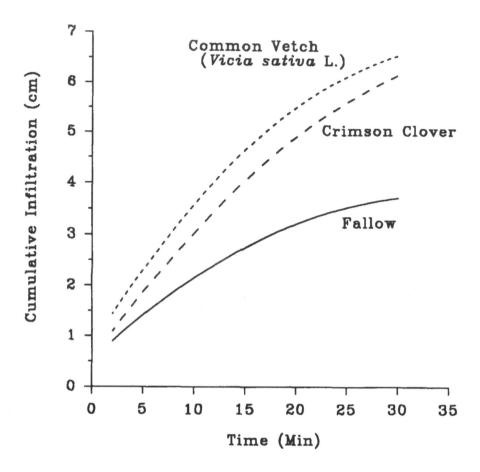

Figure 3. Effect of winter cover crop on cumulative infiltration (ring infiltrometer) following 2 years of no-tillage cotton grown on a Typic Paleudult. (From Touchton et al., 1984.)

the soil. Root fabric increases the bearing capacity of soils in no-tillage and can thus reduce the compactive effects of equipment traffic. Also, increased soil aggregation from below- and above-ground residues can improve soil structure and reduce soil bulk density and soil strength. The result of these processes alters soil strength and bulk density dependent on soil type, crop grown, and residue management system. Wagger and Denton (1989) reported a nonsignificant trend for wheat and hairy vetch cover crops to increase bulk density in the in-row position which had been subsoiled prior to planting corn. They attributed the increased bulk density, as compared to winter fallow, to the cover crops depleting soil water and interfering with soil fracturing with the subsoiler shank. A falling-plunger type penetrometer was used to measure the penetration resistance in corn, cotton, and peanut plots with and without a hairy vetch cover crop on a Coastal Plain soil in North Carolina (Lutz et al., 1946; Welch et al., 1950). In all crops, the penetration resistance (depth of penetration) was greater following the vetch cover crop and the depth of penetration was correlated to soil porosity. In contrast, researchers in Alabama found soil strength in the 6- to 18-cm depth measured at the end of the growing season was 0.3 to 0.5 MPa greater in corn plots planted behind rye and crimson clover cover crops compared to a planting behind a white lupin cover crop or winter fallow (Reeves and Touchton, 1991b). The penetrometer readings were taken when the soil was saturated after the corn had matured, so differences in soil strength between cover crops were residual in nature and were not due to differences in cover crop water use. In Nigeria, bulk density of an Alfisol increased from 1.00 g cm^{-3} under 15 years of secondary forest growth to 1.50 g cm^{-3} after 5 years of cultivation (Wilson et al., 1982). Following 5 years of cultivation, the area was cropped for 2 years using a number of grass and legume cover crops. Grass cover crops reduced bulk density to 1.33 g cm^{-3} and legume covers reduced bulk density to 1.29 g cm^{-3}. In the cotton experiment of Keisling et al. (1990) in Arkansas, bulk density in the 5- to 10-cm depth following a rye-hairy vetch cover crop averaged 1.39 g cm^{-3} compared to 1.29 g cm^{-3} without a cover crop. Bulk densities at the soil surface, and at the 10- to 15-cm depth were similar. In Alabama, neither tillage system (combinations of fall or spring disking or no-tillage) nor cover crops (rye, hairy vetch, crimson clover, or winter fallow) affected infiltration or bulk density on a Typic Paleudult cropped to cotton (Brown et al., 1985).

2. Soil Erosion Control

A cover crop by definition is sown for the purpose of erosion control (Parker, 1915). The value of soil coverage by vegetation is illustrated by the pioneering work from the Missouri Agricultural Experiment Station which showed that a continuous bluegrass (*Poa pratensis* L.) sod reduced annual erosion from 91.8 Mg ha^{-1} on bare soil to 0.7 Mg ha^{-1} with sod (Miller, 1936). This rate of loss under vegetative cover would require over 3,500 years to erode the 18 cm of surface soil on this site (Shelby loam with 3.7% slope) compared to 56 years with continuous conventional-tilled corn cultivation (Enlow and Musgrave, 1938). Cover crops reduce erosion by improving soil structure and increasing infiltration, protecting the soil surface and dissipating raindrop energy, reducing the velocity of water that moves over the soil surface (Smith et al., 1987), and by anchorage of soil by roots. The benefits of cover crops in reducing wind and water erosion and in improving productivity of eroded soils have been recently reviewed (Bruce et al., 1987; Smith et al., 1987; Langdale et al., 1991). Moreover, residue management strategies and erosion control are dealt with in another chapter of this book (see Langdale et al.).

The complementary effect of cover crops used in conjunction with conservation-tillage has been well established in recent research. The importance of the cover crop *per se* in reducing erosion relative to tillage system effects, increases as the amount of previous crop residue decreases. For example, calculated annual soil losses, based on the universal soil loss equation (USLE), of a Maury soil with a 5 % slope in Kentucky were 18 Mg ha^{-1} under conventionally tilled corn with corn residue and cover crop turned under in the spring, 2.2 Mg ha^{-1} for no-tillage without a cover crop but 6.7 Mg ha^{-1} of corn residue returned to the soil surface, and 2.0 Mg ha^{-1} for no-tillage with a winter cover crop (Frye et al., 1985). Contrast these values to those reported in a Missouri study where no-tillage corn grown for silage resulted in an annual soil loss of 22.0 Mg ha^{-1} (Wendt and Burwell, 1985). Inclusion of a rye or wheat cover crop reduced soil loss to 0.9 Mg ha^{-1}.

Compared to corn, cotton does not produce large quantities of residue. Approximately 1/3 to 1/2 of modern cotton cultivar dry matter is accumulated in stems and burs (Mullins and Burmester, 1990), the plant parts that can be returned to the soil surface and that don't decompose readily. Consequently, cotton generally produces about 2 to 3.5 Mg ha^{-1} of over-wintering residues. In Mississippi, studies on two Typic Fragiudalfs with 5 % slopes reported an annual soil loss of 74.2 Mg ha^{-1} for conventional-tilled cotton (Mutchler and McDowell, 1990). Inclusion of a wheat or hairy vetch cover crop reduced losses to 20.4 Mg ha^{-1}. No-tillage cotton averaged 19.2 Mg ha^{-1} soil loss without a cover crop and 2.3 Mg ha^{-1} with a cover crop. A winter cover crop used in combination with no-tillage or reduced tillage (no-till plant but with 3 cultivations on cotton) was the only management strategy that reduced erosion below tolerance levels (11 Mg ha^{-1} yr^{-1}). The development of improved management strategies for cover crops used in cotton production should be a high priority research focus, since the potential to reduce erosion is so great with this crop.

3. Environmental Quality

The use of cover crops to increase infiltration and reduce erosion also reduces nutrient losses in surface run-off. Nitrogen and phosphorus are the two nutrients most associated with degradation of water quality arising from agriculture (Logan, 1990). Loss of P to the environment is associated with surface runoff while loss of N is primarily through leaching losses of NO_3-N to groundwater.

Sharpley and Smith (1991) recently reviewed the effect of cover crops on surface water quality. They noted that use of a cover crop in various cropping systems consistently decreased N and P transported in runoff (Table 5). Although the amount of N and P transported in runoff is generally reduced with a cover crop, the mean annual concentration of NO_3-N and soluble P may be increased. The proportion of P in runoff that is bioavailable (soluble P plus bioavailable particulate P) may increase when cover crops are included in a residue management system. Management factors such as soil fertility level, residue management system, cover crop type, and growth stage of termination of cover crops in relation to climatological factors influence not only the amount of N and P transported in runoff but the relative proportion of these nutrients that are bioavailable. Sharpley and Smith point out that determination of bioavailable P, not just total P loading, is essential to more accurately estimate the impact of agricultural practices, including the use of cover crops, on eutrophication of surface waters (Sharpley and Smith, 1991).

During the early 1940's scientists investigated the role of cover crops to reduce losses of N via leaching. The motivation for this work was economical, i.e., more efficient use of N fertilizer. Today, concern over environmental contamination of groundwater supplies has triggered a resurgence of interest in the use of cover crops to reduce NO_3-N leaching. Recently, Meisinger et al. (1991) compiled an excellent comprehensive review of the literature concerning the ability of cover crops to reduce NO_3-N leaching. Certain facts were well established in the authors' careful review. Cover crops reduce N leaching primarily by uptake of N for production of biomass; however, reducing leachate by consumption of soil water for growth is also an important mechanism for reducing N leaching. Another factor important to a cover crop's ability to reduce N leaching is the synchronization of cover crop growth and consequent soil water and N demand with the peak leaching season, i.e., in late fall through early spring when precipitation exceeds evapotranspiration.

Averaged over the eleven studies reviewed by Meisinger et al. (1991), the effectiveness of cover crops to reduce N leaching was grasses = brassicas > legumes (Table 6). Cereal rye was very effective in reducing N leaching. This cover crop is cold tolerant, has rapid growth, and produces a large quantity of biomass. Although grasses are effective catch crops in regards to N, their wide C:N ratio results in immobilization of nitrogen, increasing the fertilizer N requirement of following crops. As Meisinger and coauthors pointed out, brassicas are equally effective in reducing the mass of N lost through the root zone; however, brassica cover crops have a much narrower C:N ratio than grass cover crops. For example, the C:N ratio for rape (*Brassica napus* L.) and radish (*Raphanus sativus* L.) reported in two studies was 20:1 or less (Muller et al., 1989; Hargrove et al., 1992). This low C:N ratio would result in substantial mineralization of N to a crop following a brassica cover crop. Brassica cover crops, however, are harder to establish, are more susceptible to plant diseases, and are not as widely adapted to a variety of climatic conditions and soils as grass cover crops. Legume cover crops, as pointed out earlier, can supply N to a following crop, but typically recover only 20 to 30% of the mass of N recovered by grass or brassica cover crops (Meisinger et al., 1991; Shipley et al., 1992).

Simulations using the EPIC (Erosion-Productivity Impact Calculator) model (Williams et al., 1984) show that the greatest potential use of cover crops to reduce NO_3 leaching impacts on water quality in the United States is in the Southeast and irrigated Midwest because of the potential for vigorous growth of cover crops in the fall and winter in these areas (Meisinger et al., 1991). These

Table 5. Effect of cover crops and residue management system (RMS) on N and P transport

Crop/RMS[a]	Cover crop	Nitrate-N[b]	Total N	Soluble P[b]	Total P	Location	Reference
		\----------------------kg ha^{-1} yr^{-1}\----------------------					
Corn/CT	None	0.36 (8.78)	0.95	0.01 (0.40)	0.15	MD	Angle et al., 1984
Corn/NT	Barley	0.04 (5.88)	0.12	0.01 (1.65)	0.01		
Corn/CT	None	2.46 (0.41)	--	0.49 (0.28)	--	NY	Klausner et al., 1974
Corn/NT	Ryegrass	1.41 (3.62)	--	0.13 (0.33)	--		
Wheat/CT	None	1.14 (0.66)	--	0.32 (0.28)	--		
Wheat/NT	Ryegrass/alfalfa	0.93 (1.26)	--	0.17 (0.23)	--		
Corn/CT	None	--	--	0.28 (0.13)	4.08	GA	Langdale et al., 1985
Corn/CT	Cereal rye	--	--	0.30 (0.20)	1.39		
Corn/CT	None	0.40 (0.81)	0.48	0.27 (0.55)	3.02	Quebec	Pesant et al., 1987
Corn/NT	Alfalfa/timothy (*Phleum pratense* L.)	0.58 (3.24)	0.59	0.24 (0.22)	0.19		
Cotton/CT	None	3.44 (3.87)	4.11	0.40 (0.43)	0.63	AL	Yoo et al., 1988
Cotton/NT	None	1.40 (1.73)	3.10	0.31 (0.39)	0.44		
Cotton/NT	Wheat	0.56 (1.12)	0.88	0.16 (0.39)	0.20		

Soybean/NT	None	3.36 (4.04)	--	0.46 (0.28)	--	MO	Zhu et al., 1989
Soybean/NT	Common chickweed (*Stelleria media* L.)	0.77 (1.86)	--	0.17 (0.45)	--		
Soybean/NT	Canada bluegrass (*Poa compressa* L.)	0.88 (1.92)	--	0.43 (0.80)	--		
Soybean/NT	Downy brome (*Bromus tectorum* L.)	0.84 (2.06)	--	0.27 (0.52)	--		
Peanut/CT	None	0.15 (0.50)	4.38	0.04 (0.14)	1.35	OK[c]	Sharpley and Smith, 1991
Peanut/CT	Ryegrass	0.08 (0.73)	1.49	0.02 (0.19)	0.47		
Peanut/CT	None	0.35 (0.29)	20.84	0.15 (0.12)	5.89	OK[c]	Sharpley and Smith, 1991
Peanut/CT	Wheat	0.19 (0.75)	3.27	0.04 (0.15)	0.92		

--, Data not available; [a]CT = Conventional tillage, NT = No-tillage; [b]Values in parentheses are mean annual concentrations in mg kg^{-1}; [c]Amounts of N and P transported in OK experiments for 6-month winter period. (From Sharpley and Smith, 1991.)

Table 6. Percentage reduction of mass of N leached for types of cover crops

Cover crop	Range	Average	Reference
Grasses	31 to 77 %	61%	Morgan et al., 1942
			Martinez and Guirard, 1990
			Karraker et al., 1950
			Meisinger et al., 1990
			Staver and Brinsfield, 1990
			Nielsen and Jensen, 1985
Brassicas	35 to 87%	62%	Chapman et al., 1949
			Volk and Bell, 1945
			Muller et al., 1989
			Bertilsson, 1988
Legumes	6 to 45%	25%	Chapman et al., 1949
			Nielsen and Jensen, 1985
			Jones, 1942
			Meisinger et al., 1990

(From Meisinger et al., 1991.)

simulations also agreed with the published literature that grasses were superior to legumes as catch crops for nitrogen.

E. Potential Disadvantages of Cover Crops

1. Establishment Costs

There are a number of potential disadvantages to using a cover crop. Fortunately, proper management techniques can decrease the negative effects of cover crops, increasing their acceptance by growers. Foremost among disadvantages is the fact that it costs money to plant the cover crop and to terminate the cover before planting the cash crop. The particular economic situation is dependent on the cash crop grown; cover crop chosen; time and method of establishment, method of termination; and the cash value applied to the environment, soil productivity and soil protection benefits derived from the cover crop. In Alabama, for example, the labor and equipment cost to plant a cover crop is estimated to be about $12.25 ha^{-1} (Crews, 1992). Using lower limits of recommended seeding rates, at the time of this writing, seed costs for a rye cover crop seeded at 67 kg ha^{-1} would be $19.76 ha^{-1}, for ryegrass seeded at 56 kg ha^{-1} $39.52 ha^{-1}, for "Tibbee" crimson clover seeded at 22 kg ha^{-1} $44.46 ha^{-1}, and for hairy vetch seeded at 34 kg ha^{-1} $41.99 ha^{-1}. In this example then, the cost of establishment of a cover crop would range from $32.01 ha^{-1} for rye up to $56.70 ha^{-1} for "Tibbee" crimson clover. The increased cost of the legume over rye can be offset by the value of the N fertilizer equivalence from the legume, 60 kg N ha^{-1} being a good estimate. It is not unreasonable to estimate that the N fertilizer requirement for a crop following a rye cover crop terminated at a late stage of growth would be increased by 25 kg N ha^{-1} due to immobilization of N by rye residue with a wide C:N ratio. Thus, the difference in cost between a rye cover crop and a legume cover crop would be offset by the value of 85 kg N ha^{-1}. At a price of $0.57 kg^{-1} for fertilizer N, this differential is worth $48.45. Ignoring considerations other than seed costs and fertilizer equivalence, legume cover crops are more profitable than grass cover crops. However, the economic risk of establishment is less with rye than with legumes due to differences in susceptibility to diseases and winter kill between rye and legumes.

Yield variance of grain crops following legume cover crops is often greater than when no cover crop is used (Allison and Ott, 1987; Franklin et al., 1989; Ott and Hargrove, 1989). Consequently, although potential profits may be higher with a legume cover crop compared to no cover crop or a grass cover crop, potential risks are also greater. This risk can be reduced by choosing an adapted legume cover crop (Franklin et al., 1989; Ott and Hargrove, 1989) and by ensuring that the cover crop is planted as early as possible to improve winter survival (Bowen et al., 1991). Economic risks and seeding costs can also be reduced by using legumes in natural reseeding cropping systems. Conservation-tillage systems that do not bury cover crop residue and seed facilitate natural reseeding. Reseeding systems can be implemented by using well-planned rotations such as that reported by Oyer and Touchton (1990). They planted crimson clover in the fall, followed that with strip-tilled soybean which was planted late enough to let the clover reseed. Corn was grown the next year in the reseeded clover (and soybean) residue, thus requiring planting of the cover crop every other year rather than annually.

In the South, grain sorghum can be planted late enough to allow crimson clover to reseed in a conservation-tillage system (Touchton et al., 1982). Corn is a more profitable crop than grain sorghum but it cannot be planted as late as grain sorghum. However, newer corn hybrids with tropically adapted germplasm can be planted later than temperate hybrids and they show great potential for being grown in conservation-tillage systems with reseeding winter annual legumes. In Alabama studies with tropical corn hybrids, equivalent or greater grain and silage yields were obtained with 50 kg N ha^{-1} in a reseeding crimson clover as with 200 kg N ha^{-1} in a fallow system (Reeves, 1992).

The introduction of legume cover crops that bloom and set seed earlier also widens the window of opportunity for achieving a reseeding system using conservation-tillage techniques. In adapted areas, crimson clover is favored over hairy vetch because it matures earlier. A new variety of crimson clover, "AU Robin", has recently been registered that sets seed 7 to 10 days earlier than "Tibbee", the earliest-maturing crimson clover cultivar previously available (van Santen et al., 1992).

Leaving 25% to 50% of the row area alive when desiccating the cover crop is another method shown to effectively allow reseeding without reducing corn grain yields (Ranells and Wagger, 1991). However, potential problems with soil water use by the strips of live cover crop during spring droughts increases the risks of yield reductions with this system (Touchton and Whitwell, 1984).

2. Soil Water Depletion

Although the residue from cover crops can increase infiltration and reduce evaporation losses, transpiration losses of soil water by the cover crop can negatively affect cash crop yields. Short term soil water depletion at the time of planting the cash crop may or may not be overcome and offset by soil water conservation later in the growing season from the cover crop residue in conservation-tillage systems, dependent on rainfall distribution in relation to crop development. In Kentucky, the early growth of corn was decreased during years of low spring rainfall due to transpiration from a hairy vetch cover crop (Corak et al., 1991). By 4 weeks after corn planting, however, soil water was conserved by the mulch from the vetch cover crop. Similar results were reported by Utomo et al. (1987) for no-till corn with both a rye and hairy vetch cover crop and by Frye and Blevins (1989) for a hairy vetch cover. Campbell et al. (1984a) reported that desiccating a rye cover crop at the time of planting a soybean crop resulted in dramatic soil water depletion by the cover crop, which delayed emergence and growth of the soybean. The reduced growth, however, resulted in greater grain yield due to more efficient water rationing by smaller soybean plants during a drought later in the season compared to conventional-tilled soybean plants following incorporation of the rye 25 days before soybean planting. Long term studies in Arkansas showed that yield decreases of cotton following plowed down cover crops of rye, hairy vetch, or rye + vetch and rye + crimson clover mixtures occurred in years characterized by a dry spring and early summer (Keisling et al., 1990).

On soils with root-restricting layers, in-row subsoiling may overcome the detrimental effect of soil water depletion by cover crops (Ewing et al., 1991) but the most universal management factor for reducing risks from early-season soil water depletion by cover crops is to desiccate the cover some time prior to planting the cash crop. Research has shown that the risk of yield reductions due to early-season depletion of soil water can be reduced by killing the cover crop 2 to 3 weeks before planting the cash crop (Wagger and Mengel, 1988; Munawar et al., 1990). On poorly drained soils, soil water depletion by the cover crop could promote an earlier planting date for the cash crop, but the practical advantage of this is probably not realistic. From a residue management standpoint, the risk for early-season soil water depletion is the same regardless of the tillage system; however, the potential for yield increases as a result of increased soil water storage and conservation later in the growing season can only be achieved in a system where cover crop residues are left on the surface.

There are a number of advantages and disadvantages to killing a cover crop early. The decision to desiccate the cover crop early must be weighed against the loss of potential benefits from a later kill date. In addition to reducing soil water depletion, killing the cover early could reduce phytotoxic effects of residues to some crops, could result in less residue to serve as the source of disease inoculum, could improve planter operation, and could improve N mineralization of nonlegume cover crops. On the other hand, killing the crop early could reduce the residue available for soil and water conservation, reduce allelopathic weed control, reduce N contribution to the following crop from legume covers and disallow potential reseeding of the cover crop. Some of these aspects have already been discussed and others will be discussed later in this chapter; however, the decision as to when to kill the cover crop must be site and situation specific.

3. Stand Reductions

Stand reductions in conservation-tillage systems following winter cover crops are frequently reported for cotton (Grisso et al., 1984; Brown et al., 1985; Hutchinson and Sharpe, 1989; Rickerl et al., 1989; Hutchinson and Shelton, 1990), and less so for corn (Campbell et al., 1984b; White and Worsham, 1989; Eckert, 1988) and soybean (Campbell et al., 1984a; Eckert, 1988). These reductions have been attributed to interference from cover crop residue with planter operations resulting in poor seed-soil contact (Mitchell and Teel, 1977; Grisso et al., 1984; Campbell et al., 1984b; Eckert, 1988), soil water depletion (Campbell et al., 1984a, 1984b; Eckert, 1988; Hutchinson and Shelton, 1990), wet soils due to residue cover (Eckert, 1988), reductions in soil temperature from residue cover (Grisso et al., 1985), allelopathic effects of cover crop residues (White et al., 1986; Hicks et al., 1989; Bradow and Bauer, 1992), increased levels of soilborne pathogens (Rickerl et al., 1986), increased predation by insects and other pests (Campbell et al., 1984b; Gaylor et al., 1984; Hutchinson and Shelton, 1990), and in the case of legume covers, un-ionized ammonia (Megie et al., 1967).

Although allelopathic effects of cover crop residue can reduce plant stands, these effects can also reduce weed populations and suppress weed growth (Shilling et al., 1984; White and Worsham, 1989; Mohler and Calloway, 1992). Lack of tillage *per se* and allelopathic effects of cover crop residue can both contribute to the suppression of weeds in conservation-tillage systems (Worsham, 1991). In addition, certain plant pathogens may be reduced by cover crops. Soil populations of *Thielaviopsis basicola*, the cause of black root rot in cotton, as well as the number of diseased cotton seedlings were reduced following hairy vetch and crimson clover cover crops compared to following winter fallow in Arkansas (Kendig and Rothrock, 1991). Although hairy vetch suppressed development of black root rot, populations of *Rhizoctonia solani* were increased following vetch compared to winter fallow (Rothrock, 1991). A better understanding of the effect of cover crops on soil ecology, including cover crop residue interaction with plant pathogens and weeds, is needed to better manage cropping systems that utilize cover crops.

Cotton is especially susceptible to stand reductions following cover crops, especially winter annual legumes. Although research suggests that legume residue may harbor higher plant pathogen populations than grass cover crops (Rothrock and Hargrove, 1988) and that cotton, compared to other crops, is especially sensitive to allelopathic effects of cover crop residues (White and Worsham, 1989; Hicks et al., 1989), lint yield reduction in cotton is not highly sensitive to stand reductions due to compensatory boll production by individual plants as plant populations decrease. This fact coupled with judicious management can reduce the risk of using cover crops in cotton to a level that is acceptable considering the advantages from using cover crops. Some research has shown that cotton cultivars vary in their sensitivity to cover crops (Hoskinson, 1984; Hicks et al., 1989; Bauer et al., 1991). Screening cultivars for sensitivity to cover crop residues would provide useful information to growers. Incorporation of cover crop residue increases the inhibitory effect to cotton seedlings (Hicks et al., 1989; White et al., 1986). Therefore, residue management systems that leave cover crop residue on the surface reduce the risk of stand reductions following winter cover crops provided the residue does not interfere with planter operation.

Probably the two most important management factors for reducing stand losses and poor growth of crops following cover crops are to desiccate the cover crop 2 to 3 weeks before planting the cash crop, and to obtain good seed/soil contact and seed placement, i.e., proper depth control. Due to budget restraints, researchers often work in small plots (which allow little room for error) and have to make do with planting equipment that is not specifically designed to work in heavy residue situations, thus the stand problems associated with equipment reported in the literature may be not as unavoidable as it would seem. Good seed placement is more challenging where residues remain on the soil surface; however, growers now have many options in planter design to facilitate planting in heavy residue. Certainly for cotton, and for corn in northern areas, equipment that removes crop residue from the immediate seeding area can help to reduce stand losses. It is well established that surface residues reduce soil temperature (van Wijk et al., 1959; Mitchell and Teel, 1977; Lal et al., 1980; Utomo et al., 1987). The relative influence of this temperature reduction on crop growth is greater in northern areas of the crop's adapted zone (van Wijk et al., 1959). Removal of residue from the zone of seed placement will not only increase soil temperature in this zone but will decrease the amount of residue that comes in contact with the seed. This results in better seed/soil contact and less allelopathic effects from residue to the developing seedling.

VI. Research and Technology Transfer Needs

The benefits of crop rotation and cover crops are well established if not well practiced in modern highly capitalized and mechanized agricultural production systems. To paraphrase an often quoted axiom of the real estate industry, "There are three areas that research on cover crops and crop rotations should address: i) *economics*, ii) *economics*, and iii) *economics*." Unless a practice is economically viable, there is no incentive for growers to adopt it. Specifically, efforts should:

i) Define the role of rotation practices and cover crops in integrated pest management (IPM) schemes. The allelopathic effects of previous crop and cover crop residues on weed ecology and crop performance need to be more closely researched. The short- and long-term agroecological interactions, especially in regard to weeds, plant pathogens, and insects, between cropping systems and residue management schemes need to be better understood. These findings have environmental as well as economic implications.

ii) Develop cropping systems that are more economically conducive to the integration of rotations and cover crops. Research areas should include reseeding cover crop systems; breeding and screening of cover crop germplasm to develop improved cover crops, e.g., improved cold tolerance of legume covers, maximization of early biomass (and in the case of legumes, N) production, and early seeding; and developing management practices and alternative rotations that improve compatibility of cover crops and crop rotation in production systems. The latter should include development of dual use cover crops and rotations, e.g., crops that serve as cash crops as well as cover crops.

iii) Develop economic risk analyses for crop rotations and cover crops. Although some work has been done in this area, more work is needed and the transfer of this information is critical for decision making by growers. We also need to develop improved management schemes that reduce the economic risk associated with the use of cover crops and rotations. Development of expert systems that facilitate the selection of cover crops and management schemes based on cover crop adaptability, hub crop selection, soil type, and climatic data would aid in managing environmental risk as well as economic risk by growers who use cover crops.

iv) Determine the economic value of the indirect, long-term or subtle benefits of rotations and cover crops, i.e., increased soil productivity, decreased erosion, and potential value in improving or maintaining environmental quality. Improve the transfer of information regarding the monetary value of these effects to growers, action agencies, and policy makers.

v) Continue research on nutrient, especially N and P, loss mechanisms and nutrient cycling in systems that integrate rotations and cover crops. Develop improved management practices for systems that use cover crops and rotations that result in more efficient use of soil water and plant nutrients. This research addresses both economic and environmental concerns.

vi) Prevent the decline of long-term tillage studies and long-term rotation studies. These studies are an invaluable source of information regarding basic soil/plant interactions as well as data base sources for economic analyses. Future studies should focus on the interactive role of rotations in residue management strategies.

References

Allison, J. R. and S. L. Ott. 1987. Economics of using legumes as a nitrogen source in conservation tillage systems. p. 145-151. In: J. F. Power (ed.) *The Role of Legumes in Conservation Tillage Systems*. Proc. of a National Conference, April 27-29, 1987, Athens, GA. Soil Conservation Society of America, Ankeny, IA.

Allmaras, R. R., G. W. Langdale, P. W. Unger, R. H. Dowdy, and D. M. van Doren. 1991. Adoption of conservation tillage and associated planting systems. p. 53-83 In: R. Lal and F. J. Pierce (eds.) *Soil Management for Sustainability*. Soil and Water Conservation Society, Ankeny, IA .

Angle, J. S., G. McClung, M. C. McIntosh, P. M. Thomas, and D. C. Wolf. 1984. Nutrient losses in runoff from conventional and no-till corn watersheds. *J. Environ. Qual.* 13:431-435.

Aulakh, M. S., J. W. Doran, D. T. Walters, A. R. Mosier, and D. D. Francis. 1991. Crop residue type and placement effects on denitrification and mineralization. *Soil Sci. Soc. Am. J.* 55:1020-1025.

Auld, D. L., B. L. Bettis, M. J. Dial, and G. A. Murray. 1982. Austrian winter and spring peas as green manure crops in northern Idaho. *Agron. J.* 74:1047-1050.

Azam, F.K., A. Malik, and M.I. Sajjad. 1985. Transformations in soil and availability to plants of ^{15}N applied as inorganic fertilizer and legume residues. *Plant and Soil* 86:3-13.

Badaruddin, M. and D.W. Meyer. 1989. Forage legume effects on soil nitrogen and grain yield, and nitrogen nutrition of wheat. *Agron. J.* 81:419-424.

Bauer, P.J., S.H. Roach, and C.C. Green. 1991. Cotton genotype response to green-manured annual legumes. *J. Proc. Agric.* 4:626-628.

Bertilsson, G. 1988. Lysimeter studies of nitrogen leaching and nitrogen balances as affected by agricultural practices. *Acta Agr. Scand.* 38:3-11.

Blevins, R.L., J.H. Herbek, and W.W. Frye. 1990. Legume cover crops as a nitrogen source for no-till corn and grain sorghum. *Agron. J.* 82:769-772.

Bowen J., L. Jordan, and D. Biehle. 1991. Economics of no-till corn planted into winter cover crops. p. 181-182 In: M.D. Mullen and B.N. Duck (eds.) *Proc. 1992 Southern Conservation Tillage Conf.*, July 21-23 1992, Jackson, TN. Spec. Pub. 92-01, Tennessee Agric. Exper. Sta., Univ. of Tennessee, Knoxville, TN.

Bradow, J.M. and P.J. Bauer. 1992. Inhibition of cotton seedling growth by soil containing LISA cover crop residues. p. 1175 In: *Proceedings Beltwide Cotton Conference*, Vol. 3. January 6-10 1992, Nashville, TN. National Cotton Council, Memphis, TN.

Brown, S.M., T. Whitwell, J.T. Touchton, and C.H. Burmester. 1985. Conservation tillage systems for cotton production. *Soil Sci. Soc. Am. J.* 49:1256-1260.

Bruce, R.R., P.F. Hendrix, and G.W. Langdale. 1991. Role of cover crops in recovery and maintenance of soil productivity. p. 109-115 In: W. L. Hargrove (ed.) *Cover Crops for Clean Water*. Proc. of an International Conference, April 9-11 1991, Jackson, TN. Soil and Water Conservation Society, Ankeny, IA.

Bruce, R.R., G.W. Langdale, and A.L. Dillard. 1990. Tillage and crop rotation effect on characteristics of a sandy surface soil. *Soil Sci. Soc. Am. J.* 54:1744-1747.

Bruce, R.R., G.W. Langdale, L.T. West, and W.P. Miller. 1992. Soil surface modification by biomass inputs affecting rainfall infiltration. *Soil Sci. Soc. Am. J.* 56:1614-1620.

Bruce, R.R., Wilkinson, S. R., and G.W. Langdale. 1987. Legume effects on soil erosion and productivity. p 127-138. In: J.F. Power (ed.) *The Role of Legumes in Conservation Tillage Systems*. Proc. of a National Conference, April 27-29, 1987, Athens, GA. Soil Conservation Society of America, Ankeny, IA.

Campbell, R.B., D.L. Karlen, and R.E. Sojka. 1984a. Conservation tillage for soybean production in the U.S. Southeastern Coastal Plain. *Soil Tillage Res.* 4:531-541.

Campbell, R.B., D.L. Karlen, and R.E. Sojka. 1984b. Conservation tillage for maize production in the U.S. Southeastern Coastal Plain. *Soil Tillage Res.* 4:511-529.

Carreker, J.R., A.R. Bertrand, C.B. Elkins, Jr., and W.E. Adams. 1968. Effect of cropping systems on soil physical properties and irrigation requirements. *Agron. J.* 60:299-302.

Carreker, J.R., S.R. Wilkinson, A.P. Barnett, and J.E. Box. 1977. *Soil and water management systems for sloping land*. ARS-S-160. U. S. Government Printing Office, Washington, D. C.

Chapman, H.D., G.F. Liebig, and D.S. Rayner. 1949. A lysimeter investigation of nitrogen gains and losses under various systems of covercropping and fertilization and a discussion of error sources. *Hilgardia* 19(3):57-95.

Conservation Technology Information Center (CTIC). 1991. *National Survey Report*. CTIC, 1220 Potter Dr., Room 170, West Lafayette, IN.

Cook, R. J. 1986. Wheat management systems in the Pacific Northwest. *Plant Disease* 70:894-898.

Corak, S.J., W.W. Frye, and M.S. Smith. 1991. Legume mulch and nitrogen fertilizer effects on soil water and corn production. *Soil Sci. Soc. Am. J.* 55:1395-1400.

Crews, J.R. 1992. *1992/93 budgets for fall/winter forage crop enterprises in Alabama*. Ala. Coop. Ext. Ser., Auburn University, Spec. Extension Rep. AEC 1-9 August 1992, 13 pp.

Crookston, R.K. and J.E. Kurle. 1989. Corn residue effect on the yield of corn and soybean grown in rotation. *Agron. J.* 82:229-232.

Daberkow, S. and M. Gill. 1989. Common crop rotations among major field crops. p. 34-40. In: *Agricultural Resources: Inputs Situation and Outlook Report*. AR-15. Economic Research Service, USDA, Washington, D. C.

Dalal, R.C., P.A. Henderson, and J.M. Glasby. 1991. Organic matter and microbial biomass in a Vertisol after 20 yr of zero-tillage. *Soil Biol. Biochem.* 23:435-441.

Decker, A.M., J.F. Holderbaum, R.F. Mulford, J.J. Meisinger, and L.R. Vough. 1987. p. 21-22 In: J.F. Power (ed.) *The Role of Legumes in Conservation Tillage Systems*. Proc. of a National Conference, April 27-29, 1987, Athens, GA. Soil Conservation Society of America, Ankeny, IA.

Dexter, A.R. 1991. Amelioration of soil by natural processes. *Soil Tillage Res.* 20:87-100.

Dicke, F.F. and W.D. Guthrie. 1988. The most important corn insects. p. 767-867. In: G.F. Sprague and J.W. Dudley (eds.) *Corn and Corn Improvement Third Edition*, ASA Monograph 18. American Society of Agronomy, Crop Science Society of America, and Soil Science Society of America, Madison, WI.

Dickson, D.W. and T.E. Hewlett. 1989. Effects of bahiagrass and nematicides on *Meloidogyne arenaria* on peanut. *Supplement to J. Nematology* 21 (4S):671-676.

Doran, J.W. 1980a. Soil microbial and biochemical changes associated with reduced tillage. *Soil Sci. Soc. Am. J.* 44:765-771.

Doran, J.W. 1980b. Microbial changes associated with residue management with reduced tillage. *Soil Sci. Soc. Am. J.* 44:518-524.

Doran, J.W. and M.S. Smith. 1991. Role of cover crops in nitrogen cycling. p. 85-90 In: W. L. Hargrove (ed.) *Cover Crops for Clean Water*. Proc. of an International Conference, April 9-11, 1991, Jackson, TN. Soil and Water Conservation Society, Ankeny, IA.

Duck, B.N. and D.D. Tyler. 1987. Soil drainage, acidity, and molybdenum effects on legume cover crops in no-till systems. p. 23-24 In: J. F. Power (ed.) *The Role of Legumes in Conservation Tillage Systems*. Proc. of a National Conference, April 27-29, 1987, Athens, GA. Soil Conservation Society of America, Ankeny, IA.

Ebelhar, S.A., W.W. Frye, and R.L. Blevins. 1984. Nitrogen from legume cover crops for no-tillage corn. *Agron. J.* 76:51-55.

Eckert, D.J. 1985. Review- effects of reduced tillage on distribution of soil pH and nutrients in soil profiles. *J. Fert. Issues* 2:86-90.

Eckert, D.J. 1988. Rye cover crops for no-tillage corn and soybean production. *J. Prod. Agric.* 1:207-210.

Edwards, J. H., D. L. Thurlow, and J. T. Eason. 1988. Influence of tillage and crop rotation on yields of corn, soybean, and wheat. *Agron. J.* 80:76-80.

Edwards, J.H., C.W. Wood, D.L. Thurlow, and M.E. Ruf. 1992. Tillage and crop rotation effects on fertility status of a Hapludult soil. *Soil Sci. Soc. Am. J.* 56:1577-1582.

Elkins, C.B., R.L. Haaland, and C.S. Hoveland. 1977. Grass roots as a tool for penetrating soil hardpans and increasing crop yields. p. 21-26. In: *Proc. 34th Southern Pasture and Forage Crop Improvement Conf., April 12-14, 1977,* Auburn University, Auburn, AL.

Enlow, C.R. and G.W. Musgrave. 1938. Grass and other thick-growing vegetation in erosion control. p. 615-645. In: *Soils and Men.* USDA Yearbook of Agriculture. U.S. Government Printing Office, Washington, D.C.

Ewing, R.P., M.G. Wagger, and H.P. Denton. 1991. Tillage and cover crop management effects on soil water and corn yield. *Soil Sci. Soc. Am. J.* 55:1081-1085.

Fleming, A.A., J.E. Giddens, and E.R. Beaty. 1981. Corn yields as related to legumes and inorganic nitrogen. *Crop Sci.* 21:977-980.

Francis, C.A. and M.D. Clegg. 1990. Crop rotations in sustainable production systems. p. 107-122. In: C. A. Edwards, R. Lal, P. Madden, R. H. Miller, and G. House (eds.) *Sustainable Agriculture Systems.* Soil and Water Conser. Soc., Ankeny, IA.

Franklin, R.L., S.L. Ott, and W.L. Hargrove. 1989. The economics of using legume cover crops as sources of nitrogen in corn production. *Faculty Series 89-11,* Division of Agricultural Economics, The University of Georgia, Athens, GA.

Froud-Williams, R.J. 1988. Changes in weed flora with different tillage and agronomic management systems. p. 213-236 In: M.A. Altieri and M. Liebman (eds.) *Weed Management in Agroecosystems.* Ecological Approaches. CRC Press, Boca Raton, FL.

Frye, W.W. and Blevins, R.L. 1989. Economically sustainable crop production with legume cover crops and conservation tillage. *J. Soil Water Conser.* 44:57-60.

Frye, W.W., R.L. Blevins, M.S. Smith, S.J. Corak, and J.J. Varco. 1988. Role of annual legume cover crops in efficient use of water and nitrogen. p. 129-154 In: W.L. Hargrove (ed.) *Cropping Strategies for Efficient Use of Water and Nitrogen.* Special Publication No. 51, American Society of Agronomy, Madison, WI.

Frye, W.W., W.G. Smith, and R.J. Williams. 1985. Economics of winter cover crops as a source of nitrogen for no-till corn. *J. Soil Water Conser.* 40:246-249.

Gardner, W.K. and D.G. Parbery. 1982. The acquisition of phosphorus by *Lupinus albus* L. I. some characteristics of the soil/root interface. *Plant and Soil* 68:19-32.

Gaylor, J.G., S.J. Fleischer, D.P. Muehlelsen, and J.V. Edelson. 1984. Insect populations in cotton produced under conservation tillage. *J. Soil Water Conser.* 39:61-64.

Gill, M. and S. Daberkow. 1991. Crop sequences among 1990 major field crops and associated farm program participation. p. 39-45 In: *Agricultural Resources: Inputs Situation and Outlook Report.* AR-24. Economic Research Service, USDA, Washington, D. C.

Gilley, J.E., J.F. Power, P.J. Reznicek, and S.C. Finkner. 1989. Surface cover provided by selected legumes. *Applied Engineering in Agriculture* 5:379-385.

Gladstones, J.S. 1976. The Mediterranean white lupin. *J. Agric. West. Aust.* 17(3):70-74.

Glínski, J. and J.L. Lipiec. 1990. *Soil physical conditions and plant roots.* CRC Press, Inc. Boca Raton, FL.

Griffith, D.R., E.J. Kladivko, J.V. Mannering, T.D. West, and S.D. Parsons. 1988. Long-term tillage and rotation effects on corn growth and yield on high and low organic matter, poorly drained soils. *Agron. J.* 80:599-605.

Grisso, R., C. Johnson, and W. Dumas. 1984. Experiences from planting cotton in various cover crops. p. 58-61 In: *Proc. Seventh Annual Southeast No-tillage Systems Conf.*, July 10, 1984, Headland, AL. Alabama Agric. Exper. Sta., Auburn University, AL.

Grisso, R.D., C.E. Johnson, and W.T. Dumas. 1985. Influence of four cover conditions in cotton production. *Trans. ASAE* 28:435-439.

Groffman, P.M., P.F. Hendrix, and D.A. Crossley, Jr. 1987. Nitrogen dynamics in conventional and no-tillage agroecosystems with inorganic fertilizer or legume nitrogen inputs. *Plant and Soil* 97:315-332.

Habib, L., J.L. Morel, A. Guckert, S. Plantureux, and C. Chenu. 1990. Influence of root exudates on soil aggregation. *Symbiosis* 9:87-91.

Hairston, J.E., J.O. Sanford, D.F. Pope, and D.A. Horneck. 1987. Soybean-wheat doublecropping: implications from straw management and supplemental nitrogen. *Agron. J.* 79:281-286.

Hanson, A.A. (ed.). 1990. *Practical Handbook of Agricultural Science.* CRC Press, Inc., Boca Raton, FL.

Hargrove, W.L. 1986. Winter legumes as a nitrogen source for no-till grain sorghum. *Agron. J.* 78:70-74.

Hargrove, W.L., J.W. Johnson, J.E. Box, Jr., and P.L. Raymer. 1992. Role of winter cover crops in reduction of nitrate leaching. p. 114-119 In: M. D. Mullen and B.N. Duck (eds.) *Proc. 1992 Southern Conservation Tillage Conf., July 21-23, 1992, Jackson, TN.* Spec. Pub. 92-01, Tennessee Agric. Exper. Sta., Univ. of Tennessee, Knoxville, TN.

Havlin, J.L., D.E. Kissel, L.D. Maddux, M.M. Claasen, and J.H. Long. 1990. Crop rotation and tillage effects on soil organic carbon and nitrogen. *Soil Sci. Soc. Am. J.* 54:448-452.

Henderson, C.W.L. 1989. Lupin as a biological plough: evidence for, and effects on wheat growth and yield. *Aust. J. Exper. Agri.* 29:99-102.

Herman, M. 1990. Effect of tillage systems and crop sequence on the rhizosphere microflora of winter wheat. *Soil Tillage Res.* 15:297-306.

Hicks, S.K., C.W. Wendt, J.R. Gannaway, and R.B. Baker. 1989. Allelopathic effects of wheat straw on cotton germination, emergence, and yield. *Crop Sci.* 1057-1061.

Holderbaum, J.F., A.M. Decker, J.J. Meisinger, F.R. Mulford, and L.R. Vough. 1990. Fall-seeded legume cover crops for no-tillage corn in the humid East. *Agron. J.* 82:117-124.

Hoskinson, P.E. 1984. Response of cotton cultivars to no-tillage in rye. p. 333. In: *Proc. Beltwide Cotton Production Research Conf., Jan. 8-12, 1984.* Atlanta, GA. National Cotton Council, Memphis, TN.

Howard, D.D. and M.D. Mullen. 1991. Evaluation of in-furrow and banded starter N P K nutrient combinations for no-tillage corn production. *J. Fertilizer Issues* 8:34-39.

Hoyt, G.D. 1987. Legumes as a green manure in conservation tillage. p. 96-98 In: J.F. Power (ed.) *The Role of Legumes in Conservation Tillage Systems. Proc. of a National Conference.* April 27-29, 1987, Athens, GA. Soil Conservation Society of America, Ankeny, IA.

Hoyt, G.D. and W.L. Hargrove. 1986. Legume cover crops for improving crop and soil management in the southern United States. *HortSci.* 21:397-402.

Huntington, T.G., J.H. Grove, and W.W. Frye. 1985. Release and recovery of nitrogen from winter annual cover crops in no-till corn production. *Commun. Soil Sci. Plant Anal.* 16:193-211.

Hutchinson, R.L. and T.R. Sharpe. 1989. A comparison of tillage systems and cover crops for cotton production on a loessial soil in northeast Louisiana. p. 517-519. In: *Proc. of the Beltwide Cotton Production Research Conf. Vol. 2, Jan 2-7, 1989*, Nashville, TN. National Cotton Council, Memphis, TN.

Hutchinson, R.L. and W.L. Shelton. 1990. Alternative tillage systems and cover crops for cotton production on the Macon Ridge. *Louisiana Agriculture* 33(4):6-8.

Johnson, N.C., P.J. Copeland, R.K. Crookston, and F.L. Pfleger. 1992. Mycorrhizae: possible explanation for yield decline with continuous corn and soybean. *Agron. J.* 84:387-390.

Jones, R.J. 1942. Nitrogen losses from Alabama soils in lysimeters as influenced by various systems of green manure crop management. *J. Am. Soc. Agron.* 34:574-585.

Karlen, D.L., E.C. Berry, and T.S. Colvin. 1991. Twelve-year tillage and crop rotation effects on yields and soil chemical properties in northeast Iowa. *Commun. Soil Sci. Plant Anal.* 19&20:1985-2003.

Karraker, P.E., C.E. Bortner, and E.N. Fergus. 1950. *Nitrogen balance in lysimeters as affected by growing Kentucky bluegrass and certain legumes separately and together.* Bull. 557. Ky. Agri. Exp. Sta., Lexington, KY.

Keisling, T.C., H.D. Scott, B.A. Waddle, W. Williams, and R.E. Frans. 1990. Effects of winter cover crops on cotton yield and selected soil properties. p. 492-496. In: *Proc. Beltwide Cotton Production Research Conference. Jan. 9-14, 1990*, Las Vegas, NV. National Cotton Council, Memphis, TN.

Kemper, B. and R. Derpsch. 1981. Results of studies made in 1978 and 1979 to control erosion by cover crops and no-tillage techniques in Paraná, Brazil. *Soil Tillage Res.* 1:253-267.

Kendig, S.M. and C.S. Rothrock. 1991. Cover crop helps cotton. Arkansas *Farm Res.* 40(3):8-9.

Kilmer, V.J. 1982. *Handbook of soils and climate in agriculture.* CRC Press, Inc., Boca Raton, Florida.

Kinloch, R.A. 1983. Influence of maize rotations on the yield of soybean grown in *Melodogyne incognita* infested soil. *J. Nematology* 15:398-405.

Kirchmann, H. 1988. Shoot and root growth and nitrogen uptake by six green manure legumes. *Acta Agric. Scand.* 38:25-31.

Klausner, S.D., P.J. Zwerman, and D.F. Ellis. 1974. Surface runoff losses of soluble nitrogen and phosphorus under two systems of soil management. *J. Environ. Qual.* 3:42-46.

Lacewell, R.D., J.G. Lee, C.W. Wendt, R.J. Lascano, and J.W. Keeling. 1989. *Crop yield and profit implications of alternative rotations and tillage practices.* Texas High Plains. Texas Agric. Exper. Stn., Bull. No. B-1619.

Ladd, J.N., M. Amato, R.B. Jackson, and J.H.A. Butler. 1983. Utilization by wheat crops of nitrogen from legume residues decomposing in soils in the field. *Soil Biol. Biochem.* 15:231-238.

Lal, R. 1976. No-tillage effects on soil properties under different crops in western Nigeria. *Soil Sci. Soc. Am. J.* 40:762-768.

Lal, R., D. De Vleeschauwer, and R.M. Nganje. 1980. Changes in properties of a newly cleared tropical alfisol as affected by mulching. *Soil Sci. Soc. Am. J.* 44:827-833.

Langdale, G.W., R.L. Blevins, D.L. Karlen, D.K. McCool, M.A. Nearing, E.L. Skidmore, A.W. Thomas, D.D. Tyler, and J.R. Williams. 1991. Cover crop effects on soil erosion by wind and water. p. 15-22. In: W.L. Hargrove (ed.) *Cover crops for clean water.* Soil and Water Conser. Soc., Ankeny, IA.

Langdale, G.W., R.A. Leonard, and A.W. Thomas. 1985. Conservation practice effects on phosphorus losses from southern Piedmont watersheds. *J. Soil Water Conser.* 40:157-160.

Langdale, G.W. and R.L. Wilson, Jr. 1987. Intensive cropping sequences to sustain conservation tillage for erosion control. *J. Soil Water Conser.* 42:352-355.

Langdale, G.W., R.L. Wilson, Jr., and R.R. Bruce. 1990. Cropping frequencies to sustain long-term conservation tillage systems. *Soil Sci. Soc. Am. J.* 54:193-198.

Leighty, C.E. 1938. Crop rotation. p. 406-430. In: *Soils and Men.* USDA Yearbook of Agriculture. U.S. Government Printing Office, Washington, D.C.

Lewis, R.D. and J.H. Hunter. 1940. The nitrogen, organic carbon, and pH of some Southeastern Coastal Plain soils as influenced by green-manure crops. *J. Amer. Soc. Agron.* 32:586-601.

Logan, T.J. 1990. Sustainable agriculture and water quality. p. 582-613. In: C.A. Edwards, R. Lal, P. Madden, R.H. Miller, and G. House (eds.) *Sustainable Agriculture Systems*. Soil and Water Conser. Soc., Ankeny, IA.

Luna, J.M. and G.J. House. 1990. Pest management in sustainable agricultural systems. p. 157-173 In: C.A. Edwards, R. Lal, P. Madden, R.H. Miller, and G. House (eds.) *Sustainable Agriculture Systems. Soil Water Conser. Soc.*, Ankeny, IA.

Lutz, J.F., W.L. Nelson, N.C. Brady, and C.E. Scarsbrook. 1946. Effects of cover crops on pore-size distribution in a Coastal Plain soil. *Soil Sci. Soc. Amer. Proc.* 11:43-46.

Mannering, J.V., L.D. Meyer, and C.B. Johnson. 1968. Effect of cropping intensity on erosion and infiltration. *Agron. J.* 60:206-209.

Martin, J.H., W.H. Leonard, and D.L. Stamp. 1976. *Principles of field crop production*. Macmillan Publishing, Co., Inc. New York, NY.

Martinez, J. and G. Guirard. 1990. A lysimeter study of the effects of a ryegrass catch crop, during a winter wheat/maize rotation, on nitrate leaching and on the following crop. *J. Soil Sci.* 41:15-16.

Mays, D.A., S.R. Wilkinson, and C.V. Cole. 1980. Phosphorus nutrition of forages. p. 805-846. In: F.E. Khasawneh, E.C. Sample, and E.J. Kamprath (eds.) *The Role of Phosphorus in Agriculture*. American Society of Agronomy, Madison, WI.

McFarland, M.L., F.M. Hons, and R.G. Lemon. 1990. Effects of tillage and cropping sequence on soil physical properties. *Soil Tillage Res.* 17:77-86.

McVay, K.A., D.E. Radcliffe, and W.L. Hargrove. 1989. Winter legume effects on soil properties and nitrogen fertilizer requirements. *Soil Sci. Soc. Am. J.* 53:1856-1862.

McVickar, M.H., E.T. Batten, E. Shulkcum, J.D. Pendleton, and J.J. Skinner. 1946. The effect of cover crops on certain physical and chemical properites of Onslow fine sandy loam. *Soil Sci. Soc. Amer. Proc.* 11:47-49.

Meek, B.D., W.R. Detar, D. Rolph, E.R. Rechel, and L.M. Carter. 1990. Infiltration rate as affected by an alfalfa and no-till cotton cropping system. *Soil Sci. Soc. Am. J.* 54:505-508.

Megie, C.A., R.W. Pearson, and A.E. Hiltbold. 1967. Toxicity of decomposing crop residues to cotton germination and seedling growth. *Agron. J.* 59:197-199.

Meisinger, J.J., W.L. Hargrove, R.L. Mikkelsen, J.R. Williams, and V.W. Benson. 1991. Effects of cover crops on groundwater quality. p. 57-68 In: W.L. Hargrove (ed.) *Cover Crops for Clean Water. Proc. of an International Conference, April 9-11, 1991.* Jackson, TN. Soil and Water Conservation Society, Ankeny, IA.

Meisinger, J.J., P.R Shipley, and A.M. Decker. 1990. Using winter cover crops to recycle nitrogen and reduce leaching. p. 3-6. In: J. P. Mueller and M. G. Wagger (eds.) *Conservation Tillage for Agriculture in the 1990's.* Spec. Bull. 90-1. N. Carolina State University, Raleigh, NC.

Meredith, H.L. 1992. Lupin fertilization studies. p. 275-276. In: *Proc. 1st European Conference on Grain Legumes.* June 1-3, 1992, Angers, France.

Miller, M.F. 1936. *Cropping systems in relation to erosion control.* Missouri Agr. Expt. Sta. Bull. 366. 36 pp.

Mills, W.C., A.W. Thomas, and G.W. Langdale. 1986. Estimating soil loss probabilities for Southern Piedmont cropping-tillage systems. *Trans. ASAE* 29:948-955.

Mitchell, W.H. and M.R. Teel. 1977. Winter-annual cover crops for no-tillage corn production. *Agron. J.* 69:569-573.

Mohler, C.L. and M.B. Calloway. 1992. Effects of tillage and mulch on the emergence and survival of weeds in sweet corn. *J. Applied Ecol.* 29:21-34.

Monroe, C.D. and E.J. Kladivko. 1987. Aggregate stability of a silt loam soil as affected by roots of maize, soybeans and wheat. *Commun. Soil Sci. and Plant Anal.* 18:1077-1087.

Morgan, M.F., H.G.M. Jacobson, and S.B. LeCompte, Jr. 1942. *Drainage water losses from a sandy soil as affected by cropping and cover crops.* Bull. 466. Conn. Agr. Exp. Sta., New Haven, CT.

Muller, J.C., D. Denys, G. Morlet, and A. Mariotti. 1989. Influence of catch crops on mineral nitrogen leaching and its subsequent plant use. p. 85-98. In: J.C. Germon (ed.) *Management Systems to Reduce Impact of Nitrates.* Elsevier Science Pub., New York, N.Y.

Mullins, G.L. and C.H. Burmester. 1990. Dry matter, nitrogen, phosphorus, and potassium accumulation by four cotton varieties. *Agron. J.* 82:729-736.

Munawar, A., R.L. Blevins, W.W. Frye, and M.R. Saul. 1990. Tillage and cover crop management for soil water conservation. *Agron. J.* 82:773-777.

Mutchler, C.K. and L.L. McDowell. 1990. Soil loss from cotton with winter cover crops. *Trans. ASAE* 33:432-436.

Neely, C.L., K.A. McVay, and W.L. Hargrove. 1987. Nitrogen contribution of winter legumes to no-till corn and grain sorghum. p. 48-49 In: J. F. Power (ed.) *The Role of Legumes in Conservation Tillage Systems.* Proc. of a National Conference, April 27-29, 1987, Athens, GA. Soil Conservation Society of America, Ankeny, IA.

Nielsen, N.E. and H.E. Jensen. 1985. Soil mineral nitrogen as affected by undersown catch crops. In: *Assessment of Nitrogen Fertilizer Requirement.* Proc. NW-European Study Ground for the Assessment of Nitrogen Fertilizer Requirement. Netherlands Fert. Inst., Haren, The Netherlands.

Norden, A.J., V.G. Perry, F.G. Martin, and J. NeSmith. 1977. Effect of age of bahiagrass sod on succeeding peanut crops. *Peanut Sci.* 4:71-74.

Norman, A. G. and C. H. Werkman. 1943. The use of the nitrogen isotope ^{15}N in determining nitrogen recovery from plant material decomposing in soil. *Agron. J.* 35:1023-1025.

Ott, S.L. and W.L. Hargrove. 1989. Profit and risks of using crimson clover and hairy vetch cover crops in no-till corn production. *Amer. J. Alter. Agric.* 4:65-70.

Oyer, L.J. and J.T. Touchton. 1990. Utilizing legume cropping systems to reduce nitrogen fertilizer requirements for conservation-tilled corn. *Agron. J.* 82:1123-1127.

Parker, E.C. 1915. *Field management and crop rotation.* Webb Publishing Co., St. Paul, MN.

Parkinson, Richard. 1799. The experienced farmer. Charles Cist, Philadelphia. In: Pieters, A. J. 1927. *Green manuring principles and practices.* John Wiley & Sons, Inc., New York, N.Y.

Pearson, L.C. 1967. *Principles of agronomy.* Reinhold Publishing Corporation, New York, N. Y.

Pendleton, J.W. 1966. Increasing water use efficiency by crop management. p. 236-258 In: W.H. Pierre, D. Kirkham, J. Pesek, and R. Shaw (eds.) *Plant Environment and Efficient Water Use.* American Society of Agronomy, Madison, WI.

Pesant, A.R., J.L. Dionne, and J. Genest. 1987. Soil and nutrient losses in surface runoff from conventional and no-till corn systems. *Can. J. Soil Sci.* 67:835-843.

Peterson, D.E. and O.G. Russ. 1982. Conservation tillage in several crop production systems. p. 11. In: *Proc. North Central Weed Control Conf., December 7-9, 1982,* Indianapolis, IN.

Peterson, T.A. and G.E. Varvel. 1989a. Crop yield as affected by rotation and nitrogen rate. III. corn. *Agron. J.* 81:735-738.

Peterson, T.A. and G.E. Varvel. 1989b. Crop yield as affected by rotation and nitrogen rate. II. grain sorghum. *Agron. J.* 81:731-734.

Pierce, F.J. and C.W. Rice. 1988. Crop rotation and its impact on efficiency of water and nitrogen use. p. 21-42. In: W.L. Hargrove (ed.) *Cropping Strategies for Efficient Use of Water and Nitrogen.* ASA Special Publication 51. ASA-CSSA-SSSA, Madison, WI.

Pieters, A.J. 1917. Green manuring: a review of the American Experiment Station literature -1. *Agron. J.* 9:62-82.

Pieters, A.J. 1927. *Green manuring principles and practices.* John Wiley & Sons, Inc., New York, N.Y.

Pieters, A.J. and R. McKee. 1938. The use of cover and green-manure crops. p. 431-444. In: *Soils and Men.* USDA Yearbook of Agriculture. U.S. Government Printing Office, Washington, D.C.

Power, J. 1991. Growth characteristics of legume cover crops in a semiarid environment. *Soil Sci. Soc. Am. J.* 55:1659-1663.

Power, J.F., J.W. Doran, and P.T. Koerner. 1991. Hairy vetch as a winter cover crop for dryland corn production. *J. Prod. Agric.* 4:62-67.

Pretty, K.M. and P.J. Stangel. 1985. Current and future use of world potassium. p. 99-128 In: R.D. Munson (ed.) *Potassium in Agriculture.* American Society of Agronomy, Madison, WI.

Ranells, N.N. and M.G. Wagger. 1991. Strip management of crimson clover as a reseeding cover crop in no-till corn. p. 174-176. In: W. L. Hargrove (ed.) *Cover Crops for Clean Water*. Proc. of an International Conference, April 9-11, 1991, Jackson, TN. Soil and Water Conservation Society, Ankeny, IA.

Reeves, D.W. 1992. Nitrogen management of tropical corn in a reseeding crimson clover conservation-tillage system. p. 11-14. In: M.D. Mullen and B.N. Duck (eds.) *Proc. of 1992 Southern Conservation Tillage Conference, July 21-23, 1992, Jackson, TN*. Spec. Pub. 92-01,Tennessee Agric. Exper. Sta., Knoxville, TN.

Reeves, D.W., J.H. Edwards, C.B. Elkins, and J.T. Touchton. 1990. In-row tillage methods for subsoil amendment and starter fertilizer application to conservation-tilled grain sorghum. *Soil Tillage Res*. 16:359-369.

Reeves, D.W. and P.L. Mask. 1992. The potential for white lupin production in the southeastern United States. p. 221-222. In: *1st European Conference on Grain Legumes*. June 1-3, 1992, Angers, France.

Reeves, D.W. and J.T. Touchton. 1991a. Deep tillage ahead of cover crop planting reduces soil compaction for following crop. *Highlights of Agricultural Res*. 38:4. Alabama Agric. Exper. Stn., Auburn University, AL.

Reeves, D.W. and J.T. Touchton. 1991b. Influence of fall tillage and cover crops on soil water and nitrogen use efficiency of corn grown on a Coastal Plain soil. p. 76-77. In: W.L. Hargrove (ed.) *Cover Crops for Clean Water. Proc. of an International Conference, April 9-11, 1991, Jackson, TN*. Soil and Water Conservation Society, Ankeny, IA.

Reeves, D.W., J.T. Touchton, and C.H. Burmester. 1986. Starter fertilizer combinations and placement for conventional and no-tillage corn. *J. Fertilizer Issues* 3:80-85.

Reeves, D.W., C.W. Wood, and J.T. Touchton. 1993. Timing nitrogen applications for corn in a winter legume conservation-tillage system. *Agron. J*. 85:98-106.

Reeves, T.G. 1984. Lupins in crop rotations. p. 207-226. In: *Proceedings of the III International Lupin Congress*. June 4-8, 1984, La Rochelle, France.

Reid, J.B. and M.J. Goss. 1982. Interactions between soil drying due to plant water use and decrease in aggregate stability caused by maize roots. *J. Soil Sci*. 33:47-53.

Reid, J.B., M.J. Goss, and P.D. Robertson. 1982. Relationship between the decrease in soil stability affected by the growth of maize roots and changes in organically bound iron and aluminum. *J. Soil Sci*. 33:397-410.

Rice, C.W. and M.S. Smith. 1984. Short-term immobilization of fertilizer nitrogen at the surface of no-till and plowed soils. *Soil Sci. Soc. Am. J*. 48:295-297.

Rickerl, D.H., W.B. Gordon, and J.T. Touchton. 1986. The effects of tillage and legume N on no-till cotton stands and soil organisms. p. 455-457. In: *Proc. Beltwide Cotton Production Research Conf., Jan. 4-9, 1986*, Las Vegas, NV. National Cotton Council, Memphis, TN.

Rickerl, D.H., W.B. Gordon, and J.T. Touchton. 1989. Influence of ammonia fertilization on cotton production in conservation tillage systems. *Commun. Soil Sci. Plant Anal.* 20(19&20):2105-2115.

Robbins, S.G. and R.D. Voss. 1991. Phosphorus and potassium stratification in conservation tillage systems. *J. Soil Water Conser.* 46:298-300.

Roder, W., S.C. Mason, M.D. Clegg, and K.R. Kniep. 1989. Crop root distribution as influenced by grain sorghum-soybean rotation and fertilization. *Soil Sci. Soc. Am. J.* 53:1464-1470.

Rodríguez-Kábana, R., D.B. Weaver, D.G. Robertson, E.L. Carden, and M.L. Peques. 1991. Additional studies on the use of bahiagrass for the management of root-knot and cyst nematodes in soybean. *Nematropica* 21:203-210.

Rodríguez-Kábana, R., C.F. Weaver, D.G. Robertson, and H. Ivey. 1988. Bahiagrass for the management of *Meloidogyne arenaria* in peanut. *Ann. Applied Nematology* 2:110-114.

Rothrock, C.S. 1987. Take-all of wheat as affected by tillage and wheat-soybean doublecropping. *Soil Biol. Biochem.* 19:307-311.

Rothrock, C.S. 1991. Cotton, cover crops and cotton seedling pathogens. p. 57-62. In: D.M. Oosterhuis (ed.) *Proc. 1991 Cotton Research Meeting, Arkansas Cotton Research/Extension/Production and Marketing Group University of Arkansas.* Arkansas Agric. Exper. Sta. Spec. Rep. 149, University of Arkansas, Fayetteville, AR.

Rothrock, C.S. and W.L. Hargrove. 1988. Influence of legume cover crops and conservation tillage on soil populations of selected fungal genera. *Can. J. Microbiol.* 34:201-206.

Rowland, I.C., M.G. Mason, and J. Hamblin. 1986. Effects of lupins on soil fertility. pp. 96-111. In: *Proc. of the 4th Int. Lupin Conf., Aug 15-22, 1986, Geraldton, Western Australia.* Western Australian Dept. of Agric., South Perth, W. A.

Russelle, M.P., O.B. Hesterman, C.C. Sheaffer, and G.H. Heichel. 1987. Estimating nitrogen and rotation effects in legume-corn rotations. p. 41-42. In: J. F. Power (ed.) *The Role of Legumes in Conservation Tillage Systems.* Proc. of a National Conference, April 27-29, 1987, Athens, GA. Soil Conservation Society of America, Ankeny, IA.

Scott, T.W., J. Mt. Pleasant, R.F. Burt, and D.J. Otis. 1987. Contributions of ground cover, dry matter, and nitrogen from intercrops and cover crops in a corn polyculture system. *Agron. J.* 79:792-798.

Sharpley, A.N. and S.J. Smith. 1991. Effects of cover crops on surface water quality. p. 41-49 In: W.L. Hargrove (ed.) *Cover Crops for Clean Water.* Proc. of an International Conference, April 9-11, 1991, Jackson, TN. Soil and Water Conservation Society, Ankeny, IA.

Sheaffer, C.C., M.P. Russelle, G.H. Heichel, M.H. Hall, and F.E. Thicke. 1991. Nonharvested forage legumes: nitrogen and dry matter yields and effects on a subsequent corn crop. *J. Prod. Agric.* 4:520-525.

Shilling, D.G., R.A. Liebl, and A.D. Worsham. 1984. Rye (*Secale cereale* L.) and wheat (*Triticum aestivum* L.) mulch: the suppression of certain broad-leaved weeds and the isolation and identification of phytotoxins. p. 244-271 In: A.C. Thompson (ed.) *The Chemistry of Allelopathy*. Proc. Am. Chem. Soc. Symp., April 1984, St. Louis, MO, American Chemical Society, Washington, D.C.

Shipley, P.R., J.J. Meisinger, and A.M. Decker. 1992. Conserving residual corn fertilizer nitrogen with winter cover crops. *Agron. J.* 84:869-876.

Skidmore, E.L., J.B. Layton, D.V. Armbrust, and M.L. Hooker. 1986. Soil physical properties as influenced by cropping and residue management. *Soil Sci. Soc. Am. J.* 50:415-419.

Smith, M.S., W.W. Frye, and J.J. Varco. 1987. Legume winter cover crops. p. 95-139. In: B.A. Stewart (ed.) *Advances in Soil Science*. Vol. 7. Springer-Verlag, New York, NY.

Somda, Z.C., P.B. Ford, and W.L. Hargrove. 1991. Decomposition and nitrogen recycling of cover crops and crop residues. p. 103-105 In: W.L. Hargrove (ed.) *Cover Crops for Clean Water*. Proc. of an International Conference, April 9-11, 1991, Jackson, TN. Soil and Water Conservation Society, Ankeny, IA .

Staver, K.W. and R.B. Brinsfield. 1990. Patterns of soil nitrate availability in corn production systems: implications for reducing groundwater contamination. *J. Soil Water Conser.* 45:318-323.

Tadano, T. and Sakai, H. 1991. Secretion of acid phosphatase by the roots of several crop species under phosphorus-deficient conditions. *Soil Sci. Plant Nutr.* 37:129-140.

Taylor, H. and L. Bull. 1992. Tillage Systems. p. 20-24. In: *Agricultural Resources: Inputs Situation and Outlook Report*. AR-25. Economic Research Service, USDA, Washington, D. C.

Torbert, H.A. and D.W. Reeves. 1991. Benefits of a winter legume cover crop to corn: rotation versus fixed-nitrogen effects. p. 99-100. In: W.L. Hargrove (ed.) *Cover Crops for Clean Water*. Proc. of an International Conference, April 9-11, 1991, Jackson, TN. Soil and Water Conservation Society, Ankeny, IA.

Touchton, J.T., W.A. Gardner, W.L. Hargrove, and R.R. Duncan. 1982. Reseeding crimson clover as a N source for no-tillage grain sorghum production. *Agron. J.* 74:283-287.

Touchton, J.T., D.H. Rickerl, C.H. Burmester, and D.W. Reeves. 1986. Starter fertilizer combinations and placement for conventional and no-tillage cotton. *J. Fertilizer Issues* 3:91-98.

Touchton, J.T., D.H. Rickerl, R.H. Walker, and C.E. Snipes. 1984. Winter legumes as a nitrogen source for no-tillage cotton. *Soil Tillage Res.* 4:391-401.

Touchton, J.T. and J.T. Sims. 1987. Tillage systems and nutrient management-
in the East and Southeast. p. 225-234. In: *Future Developments in Soil
Science Research: A Collection of SSSA Golden Anniversary Contributions
Presented at the Annual Meeting in New Orleans, LA, 30 Nov.- 5 Dec. 1986.*
Soil Science Society of America, Madison, WI.

Touchton, J.T. and T. Whitwell. 1984. Planting corn into strip-killed clover.
Highlights of Agricultural Res. 31(1):20. Alabama Agric. Exp. Sta., Auburn
University, AL.

Uhland, R.E. 1949. Physical properties of soils as modified by crops and
management. *Soil Sci. Soc. Amer. Proc.* 13:361-366.

Unger, P.W. 1984. Tillage and residue effects on wheat, sorghum, and
sunflower grown in rotation. *Soil Sci. Soc. Am. J.* 48:885-891.

USDA Plant Hardiness Map. 1990. Misc. Publication 1475, stock number 001-
000-04550-4. Superintendent of Documents, Government Printing Office,
Washington, DC 20402-9325.

Utomo, M., R.L. Blevins, and W.W. Frye. 1987. Effect of legume cover crops
and tillage on soil water, temperature, and organic matter. p. 5-6. In: J.F.
Power (ed.) *The Role of Legumes in Conservation Tillage Systems.* Proc. of
a National Conference, April 27-29, 1987, Athens, GA. Soil Conservation
Society of America, Ankeny, IA,

van Doren, D.M., Jr., W.C. Moldenhauer, and G.B. Triplett, Jr. 1984.
Influence of long-term tillage and crop rotation on water erosion. *Soil Sci.
Soc. Am. J.* 48:636-640.

van Santen, E., J.F Pedersen, and J.T. Touchton. 1992. Registration of 'AU
Robin' crimson clover. *Crop. Sci.* 32:1071-1072.

van Wijk, W.R., W.E. Larson, and W.C. Burrows. 1959. Soil temperature and
the early growth of corn from mulched and unmulched soil. *Soil Sci. Soc.
Amer. Proc.* 23:428-434.

Varco, J.J., W.W. Frye, M.S. Smith, and C.T. MacKown. 1989. Tillage
effects on nitrogen recovery by corn from a nitrogen-15 labeled legume cover
crop. *Soil Sci. Soc. Am. J.* 53:822-827.

Volk, G.M. and C.E. Bell. 1945. *Some major factors in the leaching of
calcium, potassium, sulfur and nitrogen from sandy soils: A lysimeter study.*
Bull. 416. Univ. Fla., Gainesville, FL.

von Qualen, R.H., T.S. Abney, D.M. Huber, and M.M. Schreiber. 1989.
Effects of rotation, tillage, and fumigation on premature dying of soybeans.
Plant Disease 73:740-744.

Wagger, M.G. 1989. Time of desiccation effects on plant composition and
subsequent nitrogen release from several winter annual cover crops. *Agron.
J.* 81:236-241.

Wagger, M.G. and H.P. Denton. 1989. Influence of cover crop and wheel
traffic on soil physical properties in continuous no-till corn. *Soil Sci. Soc.
Am. J.* 53:1206-1210.

Wagger, M.G. and D.B. Mengel. 1988. The role of nonleguminous cover crops in the efficient use of water and nitrogen. p. 115-127. In: W. L. Hargrove (ed.) *Cropping Strategies for Efficient Use of Water and Nitrogen*. Special Publication No. 51, American Society of Agronomy, Madison, WI.

Welch, C.D., W.L. Nelson, and B.A. Krantz. 1950. Effects of winter cover crops on soil properties and yields in a cotton-corn and cotton-peanut rotation. *Soil Sci. Soc. Amer. Proc.* 15:229-234.

Wendt, R.C. and R.E. Burwell. 1985. Runoff and soil losses for conventional, reduced, and no-till corn. *J. Soil Water Conser.* 40:450-454.

Westcott, M.P. and D.S. Mikkelsen. 1987. Comparison of organic and inorganic nitrogen sources for rice. *Agron. J.* 79:937-943.

White, R.H. and A.D. Worsham. 1989. Allelopathic potential of legume debris and aqueous extracts. *Weed Sci.* 37:674-679.

White, R.H., A.D. Worsham, and U. Blum. 1986. Control of legume cover crops for no-till and allelopathic effects. p. 412. In: *Proc. Southern Weed Science Society, 39th Annual Meeting, Jan 20-22*, Nashville, TN.

Williams, J.R., A. Jones, and P.T. Dyke. 1984. A modeling approach to determining the relationship between erosion and soil productivity. *Trans. ASAE* 27:129-144.

Wilson, D.O. and W.L. Hargrove. 1986. Release of nitrogen from crimson clover residue under two tillage systems. *Soil Sci. Soc. Am. J.* 50:1251-1254.

Wilson, G. F., R. Lal, and B. N. Okigbo. 1982. Effects of cover crops on soil structure and on yield of subsequent arable crops grown under strip tillage on an eroded Alfisol. *Soil Tillage Res.* 2:233-250.

Wischmeier, W. H. and D. D. Smith. 1978. *Predicting rainfall erosion losses- a guide to conservation planning.* Agriculture Handbook No. 537, USDA, Washington, D.C.

Wood, C.W., J.H. Edwards, and C.G. Cummins. 1991. Tillage and crop rotation effects on soil organic matter in a Typic Hapludult of northern Alabama. *J. Sustainable Agric.* 2:31-41.

Wood, C.W., D.G. Westfall, G.A. Peterson, and I.C. Burke. 1990. Impacts of cropping intensity on carbon and nitrogen mineralization under no-till dryland agroecosystems. *Agron. J.* 82:1115-1120.

Worsham, A.D. 1991. Role of cover crops in weed management and water quality. p. 141-145 In: W.L. Hargrove (ed.) *Cover Crops for Clean Water.* Proc. of an International Conference, April 9-11, 1991, Jackson, TN. Soil and Water Conservation Society, Ankeny, IA.

Yoo, K.H, J.T. Touchton, and R.H. Walker. 1988. Runoff, sediment and nutrient losses from various tillage systems of cotton. *Soil Tillage Res.* 12:13-24.

Zhu, J. C., C.J. Gantzer, S.H. Anderson, E.E. Alberts, and P.R. Beuselinck. 1989. Runoff, soil, and dissolved nutrient losses from no-till soybean with winter cover crops. *Soil Sci. Soc. Am. J.* 53:1210-1214.

Pest Management and Crop Residues

F. Forcella, D.D. Buhler, and M.E. McGiffen

I. Introduction

Crop residues affect pest populations and pest control either directly or through associated soil tillage. Because many pests and pest control techniques respond uniquely to crop residue levels and/or tillage, few definitive principles exist regarding pest management and crop residues. Nevertheless, many trends have been noted concerning this subject. Discussion of these trends will be the primary objective of this report, which has relied heavily on previous reviews by Buhler (1991), Forcella and Burnside (1992), Kirby (1985), Ruesink et al. (1989), Steffey et al. (1992), and Stinner and House (1990).

1-56670-003-5/94/$0.00+$.50

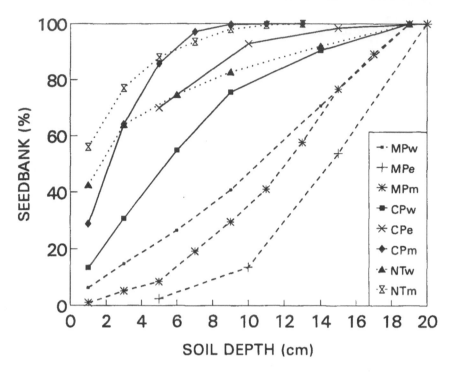

Figure 1. Depth distribution curves of weed seeds in soil after one year (thin lines) and 5 years (thick lines) of moldboard plowing, MP (dashed lines); chisel plowing, CP (solid lines); and no-tillage, NT (dotted lines) in Wisconsin (w), Minnesota (m), and England (e).

II. Tillage and Residue Incorporation

Although the individual effects of tillage and residue on pests generally have not been separated, crop residue levels typically are a function of soil tillage. Reduced tillage basically means more surface residue at the time the new crop emerges. A useful value would be an index that correlates tillage with the degree of soil mixing and residue incorporation that occurs during soil management operations. Unfortunately, quantitative tillage indices do not exist.

The depth-distribution of weed seeds in differing tillage systems permits a means for calculating a tentative tillage index. If such depth distributions are plotted (Figure 1), the relative area above each curve provides a quantitative index of soil mixing for each tillage system (Table 1). These indices are remarkably stable within tillage systems, despite the geographic disparity of the original studies: England (Cousens and Moss, 1990); Minnesota (Staricka et al., 1990, 1991); Nebraska (Wicks and Somerhalder, 1971); and Wisconsin (Yenish et al., 1992).

Table 1. Tillage index (TI) for various tillage systems after one year and five years (TI was calculated as 1 minus the normalized area under curves that describe the depth distribution of weed seeds in soil)

Tillage system	Year 1	Year 5
No-till	0.15	0.41
Ridge-till	n.a.	0.57
Chisel-plow	0.20	0.64
Spring disk	0.37	n.a.
Moldboard plow	1.12	1.01

(Original data derived from Staricka et al., 1990, 1991; Yenish et al., 1992; Cousens and Moss, 1990; and Wicks and Somerhalder, 1971.)

Theoretically, in a soil whose plow layer (0-20 cm) was thoroughly and evenly mixed, the tillage index (TI) would equal unity. Tillage systems that leave disproportionate numbers of seeds (or pest eggs, cysts, spores, or crop residue) on the soil surface would have TI values less than 1, whereas tillage systems that bury most seeds in the lower half of the plow layer would have TI values greater than 1.

The value of the TI concept is that it allows quantitative comparisons among tillage systems. Normally, tillage systems are compared graphically using histograms. For example, Figure 2a represents the density of the tillage-sensitive weed, velvetleaf (*Abutilon theophrasti*), in corn. In this graph tillage systems were arranged in their perceived order of soil disturbance (no-till, NT; ridge-till, RT; chisel-plow, CP; and moldboard plow, MP). In the first three tillage systems velvetleaf density increased in nearly equal units, but increased substantially more in MP. In contrast, when velvetleaf density was expressed as a function of TI, a distinctly linear relationship existed (Figure 2b). This relationship indicated that velvetleaf density is not so much a function of tillage type, but instead, a linear function of the degree of soil mixing associated with tillage. Accordingly, such a function allows more quantitative comparisons among tillage systems, not only for weeds, but also for important factors governing soil erosion, e.g. crop residue levels (Figure 3) and random roughness (Figure 4).

III. Weeds and Tillage Systems

Emerging generalizations regarding tillage systems and weed species are as follows: (1) annual grass and perennial weeds tend to increase in importance in reduced tillage systems, (2) relatively large-seeded broadleaf weeds tend to have higher densities in MP, and (3) small-seeded broadleaf weeds appear highly individualistic in their adaptations to tillage systems (Buhler, 1991; Forcella and Burnside, 1992).

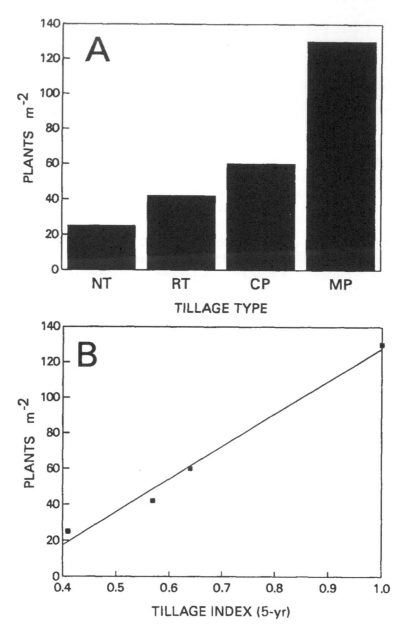

Figure 2. (a) Example of the standard manner of graphically illustrating the effect of tillage system on an agronomic variable; in this case, the density of velvetleaf. Note the relative equal differences in density among no-till (NT), ridge-till (RT), and chisel plow (CP), but the quite large increase in density in moldboard plow (MP). (b) Quantification of tillage type into a tillage index (Table 1) indicates a linear response of velvetleaf to the degree of soil mixing.

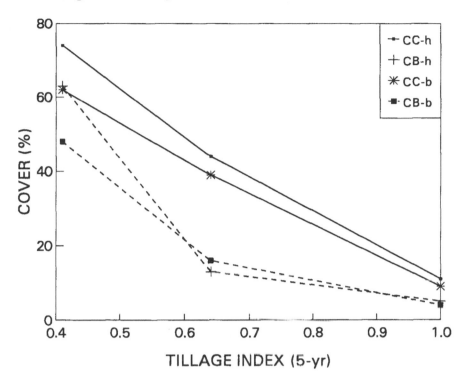

Figure 3. Relationship of tillage index to surface residue cover of corn on two soil types, Hamerly clay loam (h) and Barnes loam (b). Note that the relationship is approximately linear in continuous corn (CC), but concave in corn-soybean rotation (CB).
(Data from Lindstrom and Forcella, 1988.)

A. Weed Biology and Ecology

The increased importance of annual grasses in reduced tillage systems has been reported frequently. Examples include that of giant foxtail, *Setaria faberi* (Buhler and Oplinger, 1990; Johnson et al., 1989); green and yellow foxtail, *S. viridis* and *S. glauca* (Wrucke and Arnold, 1985; Forcella and Lindstrom, 1988); fall panicum, *Panicum dichotomiflorum*, (Williams and Wicks, 1978); and sandbur, *Cenchrus* spp., (Williams and Wicks, 1978; Wrucke and Arnold, 1985). Exceptions to this generalization occur; the tropical annual grass, itchgrass, *Rottbloellia exaltata*, apparently thrives more under MP than NT (Ayeni et al., 1984). Examples of increased densities of perennial weeds with reduced tillage include: quackgrass, *Agropyron repens* (Merivani and Wyse, 1984); Canada thistle, *Cirsium arvense* (Donald 1990a); and foxtail barley, *Hordeum jubatum* (Wrucke and Arnold, 1985; Donald, 1990b).

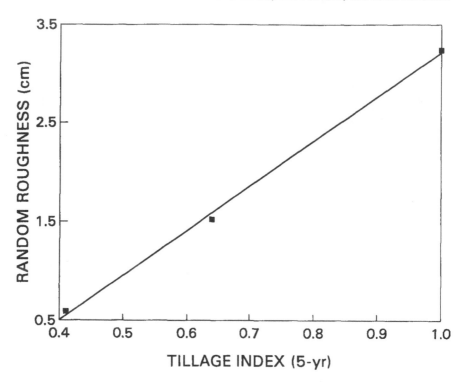

Figure 4. Relationship of tillage index to average random roughness of surface soil under MP, CP, and NT.
(Onstad and Voorhees, 1987.)

Large-seeded broadleaf weeds that decrease as tillage is reduced include cocklebur, *Xanthium strumarium* (Wrucke and Arnold, 1985); velvetleaf (Buhler and Daniel, 1988; Buhler and Oplinger, 1990) (Figure 2); and sicklepod, *Cassia obtusifolia* (Banks et al., 1985).

The variable response of small-seeded broadleaf weeds is typified by common lambsquarters, *Chenopodium album*, which Buhler and Oplinger (1990) found to increase under NT, but Putnam et al. (1983) observed to decrease under NT. Some species, such as redroot pigweed, *Amaranthus retroflexus*, increase under NT (Putnam et al., 1983), whereas the related prostrate pigweed, *A. blitoides*, decrease (Wrucke and Arnold, 1985).

In certain situations, some species appear to thrive better under moderate soil tillage (CP) rather than MP or NT. Examples include common lambsquarters (Johnson et al., 1989), redroot pigweed (Buhler and Oplinger, 1990), and velvetleaf (Freed et al., 1987). In another intermediate form of tillage, RT, green and yellow foxtail populations were found to proliferate in comparison to those in MP. However, this occurred only where corn was grown continuously.

Where corn was rotated with soybean, foxtail populations were small, manageable, and comparable to those in MP (Forcella and Lindstrom, 1988).

Although there are several reasons for differential success of weed species in tillage systems, two of the more important explanations were identified by Buhler (1991): (1) soil-depth burial preferences for germination and emergence, and (2) staggered emergence effects on herbicide efficacy.

Giant foxtail germinates and establishes its seedlings best when buried by <2 cm soil. In contrast, velvetleaf germination and seedling establishment are best when seeds are buried by 2-6 cm soil (Buhler 1991). The percent of the seedbank that is placed above 2 cm in NT, CP, and MP is about 50, 20, and 10%, respectively; whereas that placed between 2-6 cm is about 30, 20, and 20%, respectively (Figure 1). Consequently, NT positions half of the foxtail seedbank for maximum establishment and half of the velvetleaf seedbank for minimum establishment. In contrast, MP buries most foxtail seedlings too deep for effective germination, but positions about 20% of the velvetleaf seedbank at a depth that maximizes establishment. This information readily explains the differential responses of the two species to tillage, as well as the overall better weed control in MP regardless of species.

Staggered emergence of seedlings is primarily a result of seed burial depth and soil conditions that affect germination. Because surface residue acts as insulation, soils are colder and wetter in reduced tillage systems than in MP during spring. Accordingly, weeds emerge later and over a longer time period in reduced tillage systems than in MP. Consequently, efficacies of short-residual herbicides often are decreased in reduced tillage systems, especially if the herbicide was applied early preplant. Staggered emergence in reduced tillage also may lower postemergence herbicide efficacy (Buhler, 1991).

B. Control Efficacy

Crop residues may affect herbicide efficacy directly by intercepting the chemical before it reaches soil or contacts weed seedlings. However, the net effect of crop residue on weed control is somewhat contradictory (Forcella and Burnside, 1992). In some cases, crop residue retention of herbicides lowered efficacy (Erbach and Lovely, 1975); in other cases, rainfall washed intercepted herbicides into the soil and efficacy remained high (Johnson et al., 1989); and in still other instances, crop residue suppressed weed seed germination and/or seedling growth and thereby complemented the effects of herbicides (Crutchfield et al., 1986). The overall effect of crop residue on herbicide performance probably depends on many interacting factors, including specific herbicide traits and formulations, as well as pre- and post-application environmental conditions (Mills and Witt, 1989).

Tillage systems also impact the applicability of secondary tillage for weed management, i.e. rotary hoeing and interrow cultivation. Rotary hoes do not operate properly under high residue levels or rough surface conditions

(Springman et al., 1989), and therefore, they are not well suited for CT. Although interrow cultivators can be used in many CT systems, special adjustments or new heavy-duty units may be required. In RT, interrow cultivation, with its implicit ridge building, not only is an integral part of this soil management system, but also provides appreciable weed control (Forcella and Lindstrom, 1988).

There is little doubt that adoption of CT will change weed populations and alter weed management systems. Because of the central importance of herbicides in many CT systems, herbicide use may increase when tillage is reduced. Changes in application techniques and kinds of herbicides used also are likely in CT.

IV. Insects and Tillage Systems

Insects that overwinter in the soil, live in soil or plant residue, or become active during early crop growth, are the most likely insects to be affected by changes in tillage systems (Steffey et al., 1992). Generally, as tillage is reduced, the number of insect pests increases (Musick and Beasley, 1978). Many of the insects that increase in NT (Table 2) are especially troublesome (1) in continuous cereal crops, (2) when crops are sown into sod, or (3) where grass weeds are prevalent.

Reduced tillage also tends to increase diversity of predators and parasites of crop-damaging insects (Stinner and House, 1990). Regulation of pest populations by these organisms can be seen when they are removed. For example, four times more corn plants may be destroyed by black cutworms, *Agrotis ipsilon*, when predators are removed than when they are present (Stinner and House, 1990). Consequently, the switch from MP to NT often has no net effect on insect damage (Steffey et al., 1992). Indeed, previous reviews have shown that 24-29% of the pest species surveyed caused increased injury in NT, 29% were unaffected by tillage system, and 43% of the insect species examined caused less injury to corn in NT than when the same species were present in MP (Kuhlman and Steffey, 1982; Stinner and House, 1990).

A detailed example of how tillage and crop residues affect one insect, corn rootworm (*Diabrotica* spp.), is presented below, followed by brief examples for several other species.

A. Corn Rootworm

Western and northern corn rootworm (*D. vergifera* and *D. longicornis*) are the primary soil insect pests of corn in the north-central United States (Ruesink et al., 1989). Crop residue moderates soil temperatures and other microclimate factors, and affects rootworms at several points in their life cycle. Abundant crop residue in reduced tillage increases egg survivorship during cold, dry

Table 2. Examples of insect species associated with three specific tillage systems

---

Moldboard plow
    European corn borer, *Ostrinia nubilalis*
    Lesser cornstalk borer, *Elasmopalpus lignosellus*

Chisel plow
    Seedcorn maggot, *Delia platura*

No-till
    Armyworm, *Pseudaletia unipuncta*
    Black cutworm, *Agrotis ipsilon*
    Corn earworm, *Heliothis zea*
    Corn rootworm, *Diabrotica virgifera*
    Green cloverworm, *Plathypera scabra*
    Southern corn billbug, *Sphenophorus callosus*
    Stalk borer, *Papaipema nebris*

---

winters with little snow cover (Gray and Tollefson, 1988b). However, cold NT soils in spring delay initial emergence of corn rootworm adults from pupation (Gray and Tollefson, 1988a; Tyler and Eller, 1974). Corn rootworm beetles may have higher fecundities in NT than in MP (Sloderbeck and Yeargen, 1983), but most studies find no significant effects of tillage on oviposition (Levine and Oloumi-Sadeghi, 1991). While tillage can affect several aspects of the life cycle of corn rootworm, rootworm injury in NT is generally equal to (Steffey et al., 1992) or less (Musick and Beasley, 1978) than that in MP.

The most important determinant of rootworm population size in corn is crop rotation, regardless of tillage system (Steffey et al., 1992). Corn rootworm seldom causes significant crop loss when corn is rotated with other crops. Rotations are effective primarily for four reasons: (1) rootworm beetles lay eggs at the base of corn plants in summer, (2) eggs overwinter and usually hatch the following spring, (3) larvae (rootworms) are host specific and mature only on corn and a few grassy weeds (Branson and Ortman, 1970, 1971), and (4) larvae do not move readily from field to field and do not injure crops other than corn.

Although crop rotation may be the most effective tool for the management of corn rootworm, it occasionally has failed to control northern corn rootworm populations that have evolved extended diapause (Levine and Oloumi-Sadeghi, 1991). Extended diapause allows northern corn rootworm eggs to survive in the soil for two or more seasons (Krysan et al., 1986). If corn is sown in alternating years, then: (1) adult rootworm beetles may lay eggs in corn the first season, (2) the eggs can remain in diapause through the following season, and (3) then eggs hatch and larvae feed on corn roots two years hence. In Illinois, the percentage of northern corn rootworms exhibiting extended diapause varied from 14-53%; the more often corn appeared in crop rotations the higher the percentage of eggs that exhibited extended diapause (Levine et al., 1992). However, rootworm injury to corn planted the year after soybean was economically significant in only 11 of 890 Illinois fields sampled during 1986-1988. Tillage/residue effects on extended-diapause rootworm populations are unknown.

B. Other Insect Species

Crop rotational sequence often determines the composition of pests attacking crops, irrespective of tillage (Steffey et al., 1992). Rotations of corn with other crops rarely require insecticide applications.

In contrast, grass sod harbors many soil and aboveground insects that can cause serious problems in corn the following season, especially if tillage is not performed. Various wireworms and white grubs may build up large populations on grass roots and cause serious injury to corn roots the following year (Musick and Beasley, 1978). Foliage feeders, such as armyworm (*Pseudaletia unipuncta*) and the lesser stalk borer (*Elasmopalpus lignosellus*) will feed on young corn plants after grass sod has been killed (Henn et al., 1992). Rotations of clover or alfalfa with corn generally have low insect populations, but occasionally have problems with cutworms, grape colapis (*Colapis brunnea*), and seedcorn maggots (*Delia platura*) (Steffey et al., 1992).

Tillage and crop residues may affect insects indirectly by altering densities of weed species that attract some insects (e.g., cutworms, armyworms, and stalk borers) (Busching and Turpin, 1976; Musick and Petty, 1974; Gregory and Musick, 1976). Black cutworm moths lay eggs in unincorporated residue or near the base of weeds (Sherrod et al., 1979). Cutworm injury is more prevalent when early spring weed growth attracts gravid cutworm moths (Oku and Kobayashi, 1973). Stalk borers and armyworms lay eggs on grassy weeds (Levine, 1985). The larvae feed until herbicides kill the weeds; then they switch to feeding on young corn plants (Edwards and Lofty, 1978; Stinner et al., 1984). Corn planted into grass crops such as rye or pasture, or fields with grassy weeds carried over from the previous year, have an increased risk of armyworm injury (Steffey et al., 1992). Thus, increased weed populations that result from tillage decisions may increase crop losses from insect feeding as well as weed competition.

Many insect species feed in and are sheltered by crop residue. European corn borer (*Ostrinia nubilalis*) overwinters as a mature larvae in corn stalks. If the stalks are shredded or buried, fewer corn borers would be expected to survive to infest corn the following season (Musick and Petty, 1974). However, there are reports of corn borer infestations being greater in MP than in NT (Stinner and House, 1990). In soybean, CP only partially buries vegetation, leaving rotted crop residue that attracts ovipositioning seedcorn maggot (Funderburk et al., 1983). Seedcorn maggot is an interesting example of an insect that appears to prefer moderate soil disturbance (e.g., CP) in lieu of NT or MP. Crop residues in corn and soybean fields may also create a problem with stink bugs (Hemiptera: Pentatomidae). Stink bug injury is often increased if a rye cover or a wheat crop is planted before corn or soybean (Steffey et al., 1992).

C. Insecticide Efficacy

Surface residue may also affect the efficacy of soil insecticides. Soil insecticides are adsorbed to soil organic matter (Felsot and Dahm, 1979); as organic matter increases insecticides may remain bioactive longer but be less available to target pests at any given time (Felsot and Lew, 1989; Harris, 1972). Residue also may reduce efficacy by interfering with pesticide incorporation. Reports of tillage effects on insecticide behavior in soil have been contradictory (Felsot, 1987; Felsot et al., 1987; Stinner et al., 1986), and seldom have extension entomologists altered insecticide recommendations in response to changing tillage systems (Steffey et al., 1992).

In general, insect management may be somewhat more difficult in certain CT systems than MP due to specific insect problems associated with weed or cover crop vegetation, or residue from the previous crop. However, insect control in CT systems where live vegetation and crop residue have been eliminated should not require more insecticide use than in MP. Controlling weeds in the previous crop or changing the timing of weed control can reduce the need for insecticides in some CT systems. Use of integrated pest management, especially crop rotation, also can reduce insecticide use, regardless of tillage system. As a general rule, insect control problems should not prevent a farmer from adopting a conservation tillage system (Steffey et al., 1992; Musick and Beasley, 1978).

V. Other Animals and Tillage Systems

In the eastern Corn Belt, especially in wet years, slugs are a major problem in NT, but not MP. Mice consume sown crop seeds, and because mice are more prevalent in NT than in MP, NT stands are often diminished. If crop seeds are sown NT into wet soils, the planting slot frequently will not close, thereby exposing the seeds to bird depredation (Musick and Beasley, 1978).

VI. Pathogens and Tillage Systems

Tillage systems influence plant diseases by altering (1) soil water content, (2) soil temperature conditions, (3) proximity of crop and disease organism, (4) substrate (residue) availability for the pathogen, and (5) residue suitability for the pathogen. Therefore, with this variety of effects, the severity of crop diseases may increase, decrease, or remain stable with changing tillage systems (Cook et al., 1978; Kirby, 1985). In fact, not only do different pathogen species have varying tillage preferences, but even genetic races of the same disease may respond differently to tillage systems.

A. Soil Water and Temperature

The case of the take-all fungus (*Gaeumannomyces graminis*) illustrates the complicated relationship among tillage, water, temperature, and plant disease. Take-all requires high soil temperatures and soil water contents for infection. Consequently, colonization of seedlings of spring cereals may be greater in MP than NT because the relatively low soil temperatures in NT inhibit infection. In contrast, seedlings of winter cereals are more at risk in NT than MP because the relatively high soil water content at planting (late summer) in NT promotes infection (Cook et al., 1978). Similarly, corn root rot (*Phythium ultimum*) is present in most corn fields, but generally does not injure plants unless soils are cold and wet. Reduced tillage often lowers soil temperature and increases soil water, inducing increased corn root rot (van Wijk et al., 1959).

B. Proximity of Crop and Disease Organism

Crop residue is a prime substrate for many disease organisms. In the absence of tillage, crop residue and soil-borne pathogens are concentrated in the upper 15 cm of soil (Sumner et al., 1981). Tillage redistributes residue and disease organisms throughout the plow layer, thereby decreasing the inoculum level. Consequently, most pathogens that infect above-ground crop tissues decrease in importance with tillage (Boosalis and Doupnik, 1976). Leaf blight and bacteria are good examples of these effects. Fungal leaf blights (*Helminthosporium* spp.) depend entirely upon aboveground residue for over-winter survival. Pathogenic bacteria are generally non-motile, thus wind and water are their main forms of transport from residue to living crop plants. Accordingly, NT crops tend to have more bacterial and leaf blight diseases than MP crops. In contrast, root diseases also depend upon crop residues for survival, but residue burial by tillage does not control root diseases.

C. Substrate Availability and Suitability

Most above-ground pathogens do not colonize residue already occupied by other microbial colonists (Cook et al., 1978). For example, if wheat straw is left on the soil surface to weather before being buried, it is colonized by air-borne saprophytes. This process preempts the substrate, inhibits colonization by the causal agents of root, crown, and foot disease of wheat, and reduces crop damage.

Some crop residues reduce pathogen damage to crops. There may be several mechanisms for this phenomenon: (1) inhibitory chemicals may leach from decomposing residue, (2) stimulatory chemicals leach from residues and promote populations of beneficial microbial control agents, (3) high C:N ratios enhance populations of highly competitive non-pathogenic species in lieu of non-competitive pathogenic species, and (4) higher soil water contents increase vigor of crops making them less susceptible to diseases.

Since the primary methods of disease control in most agronomic crops continue to be crop rotation and genetic resistance (i.e., substrate availability and suitability, respectively), changes in disease incidence due to tillage may not impact pesticide use greatly. Fungicide seed-treatments are used almost universally in corn, but only seldomly in most other crops. Nevertheless, because of the increased likelihood of seedling damping off and root diseases with NT, use of fungicide seed-treatments might be expected to increase in NT systems.

VII. Summary and Conclusions

The intensive tillage traditionally conducted in row crop production has a profound effect on pests and their control. Effective pest management in conservation tillage systems will require integration of new information with established principles of pest management. New technologies must be developed to deal with the altered ecosystems created by conservation tillage practices. Current knowledge indicates that many pests behave differently in these new systems. This new information must be integrated with the long-standing principles of pest management to develop economically and environmentally sound pest management systems. It is the pest manager's challenge to develop management systems based on the biological idiosyncrasies of new interactions among pests, crops, residues, and the environment.

References

Ayeni, A.O., W.B. Duke, and I.O. Akobundu. 1984. Weed interference in maize, cowpea, and maize/cowpea intercrop in a subhumid tropical environment. I. Influence of cropping season. *Weed Res.* 24:249-279.

Banks, P.A., T.N. Tripp, J.W. Wells, and J.E. Hammel. 1985. Effects of tillage on sicklepod (*Cassia obtusifolia*) interference with soybeans (*Glycine max*) and soil water use. *Weed Sci.* 34:143-149.

Boosalis, M.G. and B. Doupnik. 1976. Management of crop residues in reduced tillage systems. *Entomol. Soc. Am. Bull.* 22:300-302.

Branson, T.F. and E.E. Ortman. 1970. The host range of the larvae of the western corn rootworm: Further studies. *J. Econ. Entomol.* 63:800-803.

Branson, T.F. and E.E. Ortman. 1971. The host range of the larvae of the western corn rootworm: Further studies. *J. Kansas Entomol. Soc.* 44:50-52.

Buhler, D.D. 1991. Influence of tillage systems on weed population dynamics and control in the northern Corn Belt of the United States. *Trends in Agronomy* 1:51-59.

Buhler, D.D. and T.C. Daniel. 1988. Influence of tillage systems on giant foxtail, *Setaria faberi*, and velvetleaf, *Abutilon theophrasti*, density and control in corn, *Zea mays* L. *Weed Sci.* 36:642-647.

Buhler, D.D. and E.S. Oplinger. 1990. Influence of tillage systems on annual weed densities and control in solid-seeded soybean (*Glycine max*). *Weed Sci.* 38:158-165.

Busching, M.K. and F.T. Turpin. 1976. Oviposition preferences of the black cutworm moths among various crop plants, weeds, and plant debris. *J. Econ. Entomol.* 69:587-590.

Cook, R.J., M.G. Boosalis, and B. Doupnik. 1978. Influence of crop residues on plant diseases. p. 147-163. In: W.R. Oshwald (ed.) *Crop Residue Management Systems.* ASA Special Publication 31, Madison, WI.

Cousens, R. and S.R. Moss. 1990. A model of the effects of cultivation on the vertical distribution of weed seeds within the soil. *Weed Res.* 30:61-70.

Crutchfield, D.A., G.A. Wicks, and O.C. Burnside. 1986. Effect of winter wheat (*Triticum aestivum*) straw mulch level on weed control. *Weed Sci.* 34:110-114.

Donald, W.W. 1990a. Management and control of Canada thistle, *Cirsium arvense*. *Reviews of Weed Sci.* 5:193-249.

Donald, W.W. 1990b. Primary tillage for foxtail barley (*Hordeum jubatum*) control. *Weed Tech.* 4:318-321.

Edwards, C.A. and J.R. Lofty. 1978. The influence of arthropods and earthworms upon root growth of direct drilled cereals. *J. Applied Ecology* 15:789-795.

Erbach, D.C. and W.G. Lovely. 1975. Effect of plant residue on herbicide performance in no-tillage corn. *Weed Sci.* 23:512-515.

Felsot, A.S. 1987. Fate and interactions of pesticides in conservation tillage systems. p. 35-43. In: G.J. House and B.R. Stinner (eds.) *Arthropods in Conservation Tillage Systems.* Entomol. Soc. Amer. College Park, MD.

Felsot, A.S. and P.A. Dahm. 1979. Sorption of organophosphorus and carbamate insecticides by soil. *J. Agricul. Food Chem.* 27:557-563.

Felsot, A.S. and A. Lew. 1989. Factors affecting bioactivity of soil insecticides: relationships among uptake, desorption, and toxicity of carbofuran and terbufos. *J. Econ. Entomol.* 82:389-395.

Felsot, A.S., W.N. Bruce, and K.L. Steffey. 1987. Degradation of terbufos (Counter) soil insecticide in corn fields under conservation tillage practices. *Bull. Environmental Contamination Toxicology* 38:369-376.

Forcella, F. and M.J. Lindstrom. 1988. Weed seed populations in ridge and conventional tillage. *Weed Sci.* 36:500-502.

Forcella, F. and O.C. Burnside. 1992. Pest management - weeds. In: J. Hatfield and D. Karlen (eds.) *Sustainable Agriculture: The New Conventional Agriculture.* CRC Press, Boca Raton, FL. In press.

Freed, B.E., E.S. Oplinger, and D.D. Buhler. 1987. Velvetleaf control for solid-seeded soybean in three corn residue management systems. *Agron. J.* 79:119-123.

Funderburk, J.E., L.P. Pedigo, and E.C. Berry. 1983. Seedcorn maggot (Diptera: Anthomyiidae) emergence in conventional and reduced tillage soybean systems in Iowa. *J. Econ. Entomol.* 76:131-134.

Gray, M.E. and J.J. Tollefson. 1988a. Emergence of the western and northern corn rootworms (Coleoptera: Chrysomelidae) from four tillage systems. *J. Kansas Entomol. Soc.* 61:186-194.

Gray, M.E. and J.J. Tollefson. 1988b. Influence of tillage systems on egg populations of western and northern corn rootworms (Coleoptera: Chrysomelidae). *J. Kansas Entomol. Soc.* 61:186-194.

Gregory, W.W. and G.J. Musick. 1976. Insect management in reduced tillage systems. *Bull. Entomol. Soc. Amer.* 22:302-304.

Harris, C.R. 1972. Factors influencing the effectiveness of soil insecticides. *Annual Rev. Entomol.* 17:177-198.

Henn, T., Weinzierl, R.A., Gray, M.E., and Steffey, K.L. 1992. Alternatives in insect management: field and forage crops. p. 51-75. In: *Illinois Pest Control Handbook.* Univ. of Illinois Coop. Ext. Serv., Urbana, Illinois.

Johnson, M.D., D.L. Wyse, and W.E. Lueschen. 1989. The influence of herbicide formulation on weed control in four tillage systems. *Weed Sci.* 37:239-249.

Kirby, H.W. 1985. Conservation tillage and plant disease. p. 131-136. In: F.M. D'Itri (ed.) *A Systems Approach to Conservation Tillage.* Lewis Publishers, Chelsea, MI.

Krysan, J.L., D.E. Foster, T.F. Branson, K.R. Ostlie, and W.S. Cranshaw. 1986. Two years before the hatch: rootworms adapt to crop rotation. *Bull. Entomol. Soc. Am.* 32:250-253.

Kuhlman, D.E. and K.L. Steffey. 1982. Insect control in no-till corn. p. 118-147. *Proceedings 37th Annual Corn and Sorghum Research Conference.* American Seed Trade Association, Washington, D.C.

Levine, E. 1985. Oviposition by the stalk borer, *Papaipema nebris* (Lepidoptera: Noctuidae) on weeds, plant debris and cover crops in cage tests. *J. Econ. Entomol.* 78:65-68.

Levine, E. and H. Oloumi-Sadeghi. 1991. Management of Diabroticite rootworms in corn. *Annual Rev. Entomol.* 36:229-255.

Levine, E., H. Oloumi-Sadeghi, and J.R. Fisher. 1992. Discovery of multiyear diapause in Illinois and South Dakota northern corn rootworm (Coleoptera: Chrysomelidae) eggs and incidence of the prolonged diapause trait in Illinois. *J. Econ. Entomol.* 85:262-267.

Lindstrom, M.J. and F. Forcella. 1988. Tillage and residue management effects on crop production in the northwestern Corn Belt. p. 565-567. In: P.W. Unger et al., (eds.) *Challenges in Dryland Agriculture.* Texas Agric. Exp. Stn., Amarillo.

Merivani, Y.N. and D.L. Wyse. 1984. Effects of tillage and herbicides on quackgrass in a corn-soybean rotation. *Proceedings North Central Weed Control Conference* 39:97.

Mills, J.A. and W.W. Witt. 1989. Effect of tillage systems on the efficacy and phytotoxicity of imazaquin and imazethapyr in soybean (*Glycine max*). *Weed Sci.* 37:233-238.

Musick, G.J. and L.E. Beasley. 1978. Effect of crop residue management system on pest problems in field corn (*Zea mays* L.) production. p. 173-186. In: W.R. Oshwald (ed.) *Crop Residue Management Systems.* ASA Special Publication 31, Madison, WI.

Musick, G.J. and D.L. Collins. 1971. Northern corn rootworm affected by tillage. *Ohio Rep.* 56:88-91.

Musick, G.J. and H.B. Petty. 1974. Insect control in conservation tillage systems. p. 47-52. In: *Conservation Tillage.* Soil Conserv. Soc. Am. Ankeny, IA.

Oku, T. and T. Kobayashi. 1973. Studies on the ecology and control of insects in grasslands. V. Oviposition behavior of the black cutworm moth, *Agrotis ypsilon* Hufnagel, with notes on some larval behaviors. *Bull. Tohoku National Agricultural Experiment Station* 46:161-183.

Onstad, C.A. and W.B. Voorhees. 1987. Hydrologic soil properties affected by tillage. p. 95-112. In: T.J. Logan, et al. (eds.) *Effects of Conservation Tillage on Groundwater Quality: Nitrates and Pesticides.* Lewis Publishers, Chelsea, MI.

Putnam, A.R., J. DeFrank, and J.B. Barnes. 1983. Exploitation of allelopathy for weed control in annual and perennial cropping systems. *J. Chem. Ecol.* 9:1001-1010.

Ruesink, W.G., H. Oloumi-Sadeghi, S.M. Arif, D.J. Fielding, M.E. McGiffen, and W.O. Lamp. 1989. *Research needs in corn pest management in the north central United States.* N.C. Regional Committee NCS-3, Purdue University, W. Lafayette, IN. 41 p.

Sherrod, D.W., J.T. Shaw, and W.H. Luckmann. 1979. Concepts on black cutworm field biology in Illinois. *Environ. Entomol.* 8:191-195.

Sloderbeck, P.E. and K.V. Yeargen. 1983. Green cloverworm (Lepidoptera: Noctuidae) populations in conventional and double-cropped no-till soybeans. *J. Econ. Entomol.* 76:785-791.

Springman, R., D. Buhler, R. Schuler, D. Mueller, and J. Doll. 1989. *Row crop cultivators for conservation tillage systems.* Univ. Wisconsin Extension Publ. A-3483.

Staricka, J.A., P.M. Burford, R.R. Allmaras, and W.W. Nelson. 1990. Tracing the vertical distribution of simulated shattered seeds as related to tillage. *Agron. J.* 82:1131-1134.

Staricka, J.A., R.R. Allmaras, and W.W. Nelson. 1991. Spatial variation of crop residue incorporated by tillage. *Soil Sci. Soc. Am. J.* 55:1668-1674.

Steffey, K.L., M.E. Gray, and R.A. Weinzierl. 1992. *Insect management in conservation tillage systems. Midwest Plan Service.* Ames, IA. (in press).

Stinner, B.R. and G.J. House. 1990. Arthropods and other invertebrates in conservation-tillage agriculture. *Annual Rev. Entomol.* 35:299-318.

Stinner, B.R., H.R. Krueger, and D.A. McCartney. 1986. Insecticide and tillage effects on pest and non-pest arthropods in corn agroecosystems. *Agriculture, Ecosystems and Environment* 15:11-21.

Stinner, B.R., D.A. McCartney, and W.A. Rubnick. 1984. Some observations on the ecology of the stalk borer (*Papaipema nebris* (GN): Noctuidae) in no-tillage agroecosystems. *J. Georgia Entomol. Soc.* 19:229-234.

Sumner, D.R., B. Doupnik, Jr., and M.G. Boosalis. 1981. Effects of tillage and multiple cropping on plant diseases. *Annual Rev. Phytopathology* 19:167-187.

Tyler, B.M.J. and C.R. Eller. 1974. Adult emergence, oviposition and lodging damage of northern corn rootworm (Coleoptera: Chrysomelidae) under three tillage systems. *Proceedings Entomological Society Ontario* 105:86-89.

van Wijk, W.R., W.E. Larson, and W.C. Burrows. 1959. Soil temperature and the early growth of corn from mulched and unmulched soil. *Soil Sci. Soc. Am. Proc.* 23:428-434.

Wicks, G.A. and B.R. Somerhalder. 1971. Effect of seedbed preparation for corn on distribution of weed seed. *Weed Sci.* 19:666-668.

Williams, J.L. and G.A. Wicks. 1978. Weed control problems associated with crop residue systems. p. 165-172. In: W.R. Oshwald (ed.) *Crop Residue Management Systems.* ASA Special Publication 31, Madison, WI.

Wrucke, M.A. and W.E. Arnold. 1985. Weed species distribution as influenced by tillage and herbicides. *Weed Sci.* 33:853-856.

Yenish, J.P., J.D. Doll, and D.D. Buhler. 1992. Effects of tillage on vertical distribution and viability of weed seed in soil. *Weed Sci.* 40:429-433.

Crop Residue Management: Soil, Crop, Climate Interactions

D.C. Reicosky

I. Introduction

Our agricultural production system is under increasing pressure to provide low cost, high quality food and fiber while maintaining and preserving the environment. Potential environmental hazards of modern agricultural practices, decreasing profit margins, and rapidly degrading soil and water resources have been and continue to be a major concern. With increasing population pressures and increased intensity of agricultural production, additional efforts are needed to maintain the sustainability of our production systems. Increased interest in sustainability is related to the recognition that the soil, water and air resources are finite. There is also a sense of urgency about establishing improved residue management practices to meet current congressional mandates to reduce soil erosion. Accelerated communication is essential for getting the farmers acquainted with and skilled in available crop residue management technology in time to avoid the penalties associated with noncompliance. The trade-offs between economic food production and environmental quality must be addressed and require better political and technical decisions to maintain sustainability.

One of the key factors for maintaining soil and water resources is to minimize wind and water erosion that decreases productivity. Emphasis has been placed on reducing soil erosion through use of conservation production systems with improved crop residue management. Improved crop residue management can significantly reduce soil erosion and runoff, enhance moisture retention, lower summer soil temperatures, reduce the trips across the field, reduce machinery costs and at the same time may increase the net return to the farmer. Erosion must be significantly reduced on many of our soils before soil productivity drops so low that the crop production is no longer economically sustainable. The basic scientific principles should work anywhere when we utilize site-specific soil, crop and climate information and integrate this information to make intelligent decisions for managing the crop for maximum residue production. We no longer have the luxury of being able to make broad generalizations without having data for site-specific recommendations.

Considerable research has been conducted on residue management in two major problem areas: water conservation in the semi-arid areas and soil erosion in the humid areas. Water storage in the soil is essential because plants use ample water between precipitation or irrigation events. In the Great Plains where rainfall is often limited, much of the residue management research is related to water conservation. Efforts are included for snow trapping (Snyder et al., 1979; Smika and Whitfield, 1966; Benoit et al., 1986) and infiltration enhancement for greater water storage. Excellent comprehensive reviews have been presented by Unger et al. (1988) and Steiner (1992), on the role of crop residues in improving water conservation and is beyond the scope of this chapter.

The major goal of residue management in humid regions of the eastern United States is for water erosion control. Rainfall amounts and high intensity thunderstorms can result in substantial soil movement. Soil erosion research was reviewed by Wischmeier (1973) and the Soil Conservation Service (1967). Edwards and Owens (1991) have demonstrated the importance of high intensity storms on total erosion. They found with 4000 rainfall events during a 28-year study period, the five largest erosion producing events accounted for 66% of the total erosion. More than 92% of the erosion came in corn years of a corn-wheat-meadow-meadow rotation. Residue management for the long term must control erosion from intermittent severe storms, especially during susceptible crop growth periods.

One complicating aspect of residue management in the United States is the several physiographic regions related to different climate, crop and soil interactions. The wide variation in climate and production systems precludes general recommendations for all parts of the country. A large number of crop species must be integrated into a number of cropping systems to achieve the desired production. Thus there are countless combinations of crops, cover crops and cropping systems for erosion control that need to be considered in view of the soil, climate, and crop interactions. In many cases the dual role of the "dead" crop residues is vital for erosion control and becomes more important in the role of "live" crop residues, i.e. cover crops, that are used for both erosion control and nutrient scavengers (Power and Biederbeck, 1991).

The objective of this work is to present an overview of the complex interaction between soil, crop and climate as they affect residue production and management under a variety of soil-crop-climate regimes across the United States. We need enhanced understanding of crop residue management from a site-specific scale to a national scale and to identify knowledge gaps that need further research to enhance technology transfer. This information will allow farmers to make intelligent decisions for enhanced and efficient food production while maintaining environmental quality.

II. Soils

The soils across the United States vary in soil type, depth, parent material, topography and many physical, chemical and biological characteristics that result in over 12,000 different mapped soils. This large number of soils and confounding soil physical properties related to clay type, texture, organic matter content and the inherent natural fertility presents a complex and variable medium for crop residue management. Much quantitative information is available in soil data bases, e.g., the SCS Soils V data base with Soil Interpretation Records (SIR) that contain physical and chemical data useful in making crop management decisions, information for construction and water management. The land capability classes are described and recommendations for potential plant growth from the native plant community are included. While much of this information is very helpful in making management decisions, the Soils V data base does not include unsaturated hydraulic conductivity and only limited water retention data over the range experienced by actively growing plants. This data provides a starting point for further modification and enhanced quantification of parameters critical for use in site-specific recommendations. The data provides only a range of specific parameters and little assistance on how to extrapolate or interpolate between points to utilize the information on a broader scale.

The interaction between the soil depth available for rooting and the potential plant rooting depth is not well understood. Plant roots are important for extracting nutrients and water to enable the plants to survive short-term drought and generate sufficient biomass for residue management. Readers are referred to recent reviews by Brown and Scott (1984), Hamblin (1985) and McCoy (1987) for detailed discussions of soil physical effects on root growth. Borg and Grimes (1986) and Doorenbos and Pruitt (1977) list the expected maximum rooting depth for several crop species under "favorable environmental" conditions that range from 0.4 to 6.0 m deep depending on the species. This type of information is of limited value when we recognize the primary limiting factor for rooting depth in the field is often soil strength, bulk density, aeration and water availability. Low soil temperatures or low water holding capacity can also restrict root depth. A specific example is illustrated in Table 1 from Fehrenbacher and Rust (1956) who measured corn (*Zea mays* L.) rooting depth in four Indiana soils with different subsoil physical properties. The longtime average corn yields were related to root development and available water. The primary reason given by the authors for increased yields was the improved availability of water.

A second example of limiting soil physical properties is illustrated in the work by Tennant (1976) in Table 2. The same variety of spring wheat grown at one

Table 1. Relation of effective rooting depth to average corn yield on four Indiana soils

Soil series	Approx. effective root depth (m)	Available water (mm)	Average yield (kg ha^{-1})
Saybrook	1.37	269	4830
Ringwood	1.22	249	4704
Elliott	0.91	180	4077
Clarence	0.91	163	3763

(Adapted from Fehrenbacher and Rust, 1956.)

Table 2. Maximum depth of spring wheat roots on four soil types at Tammin, Western Australia

Soil type	1969	1970	1971
	--------Depth (m)--------		
1. Uniform deep loamy sand	1.40	1.69	1.65
2. Sandy loam, 14% clay increasing to 23%	1.58	1.73	1.68
3. Gray cracking clay: calcrete at 30-45 cm	0.26	0.31	0.28
4. Sand over gray clay (10-25 cm sand)	0.61	0.73	0.70
Growing season rain (mm)	126	223	165

(Adapted from Tennant, 1976.)

location on four different soil types resulted in different rooting depths determined primarily by soil physical properties regardless of rainfall in each of the three years. These results confirm the importance of the interaction of limiting soil physical factors and rainfall and other chemical characteristics that affect rooting depth regardless of the crop potential.

For soils not adequately protected by crop residues, soil erodibility is often the largest determinant of soil erosion. Soil erodibility is the ease with which soil is detached by splash during rainfall and/or by shear of surface water flow. Young et al. (1990) have shown that the concept of a constant soil erodibility factor is no longer valid and that soil erodibility is dynamic with time, soil moisture and temperature, soil disturbance due to tillage or animal activity and biological and chemical factors. Variations in the soil erodibility through the season were primarily correlated with three factors: (1) soil temperature that includes freezing and thawing, (2) soil texture, and (3) antecedent soil water. Adequate characterization of soil erodibility requires a dynamic representation of the complex interaction of soil physical and chemical properties and climate related to antecedent soil moisture and freezing and thawing.

In Northern climatic zones, surface soil freezing and thawing can have a significant effect on the soil erodibility by reducing bulk density, hydraulic conductivity, shear strength and aggregate stability (Benoit et al., 1986; Benoit, 1973). Freezing and thawing decreases the water stable aggregates and results in substantially more erosion. Low soil bulk density and high soil water content shortly after thawing provide a soil surface very susceptible to soil detachment and transport and when combined with high intensity spring rains or snowmelt often results in large soil losses. Hence the need for adequate residue cover to prevent such soil losses before crop canopy closure.

In the northern U.S., soil freezing can occur to depths greater than 1 m. The physical processes of soil freezing can be altered indirectly by residue management. Benoit et al. (1986) have shown standing residue and available snow trapped by the residue can decrease the depth of frozen soil. They compared no residue with chopped residue and standing residue and showed an insulating effect of snow trapped in the standing residue kept the soil from freezing below a depth of 0.5 m compared to 1.1 m on the no residue treatment. There was a strong inverse linear relationship between soil frost depth and snow depth accumulated prior to the cold period. Little information is available on the effect of depth of frozen soil and subsequent effects on chemical movement and groundwater quality. Freeze-thaw cycles at the soil surface have a significant effect on soil erosion and require improved residue management for minimum soil loss. Deeper frozen soils with fewer freeze-thaw cycles may require different residue management to minimize impacts on groundwater quality.

III. Crops

Crops are usually selected for maximum grain yield and economic return and generally not selected for total amount or type of residue. The crop is the primary determinant of amount and quality of residue left to protect and maintain the soil productivity. The adaptability of the various crops is often limited by climate. Over 100 major species are used in U.S. agriculture for various types of food production. A system for improved residue management will require an enhanced data base for a complete understanding of these many species and their soil and climatic adaption.

One of the keys to long-term residue management is to understand that surface residue is a precursor to soil organic matter. Factors that affect the amount and duration of organic matter in the soil will also have a direct bearing on residue duration on the soil surface. MacRae and Mehuys (1985) have reviewed the effects of cover crops and other factors on soil organic matter and soil physical properties in the temperate soils of the U.S. summarized in Table 3. The primary factors are the climate, the soil and the quantity and composition of plant residue. Management practices have a direct bearing on the maintenance of organic matter and need to be carefully considered in any residue management system.

Table 3. Factors influencing the maintenance and accumulation of organic matter in soil

Climate
Temperature, solar radiation
Precipitation, evaporation
Soil and site
Elevation, slope, aspect, geographical location
Soil type
Texture
Structure, compaction
Native organic matter and humus content
Soil temperature
Soil moisture, aeration
pH
Mineral ion content
Plant cover-species, density, distribution, history of site
Microbial populations-species, density, distribution, history of site
Faunal populations-species, density, distribution, history of site
Use of fertilizers, lime, mulches, and pesticides
Tillage, cultivation, drainage, irrigation
Fire, e.g., burning of crop residues
Material incorporated
Composition (e.g., carbohydrates, proteins, lignins, fats, waxes)
Quantity added per unit area
Moisture content
C:N ratio
Mineral ion content
Timing, method, and frequency of incorporation
Experimental variables
Plot size, arrangement, number of replicates
Cultivation practices
Frequency of sampling
Sample size, shape
Technique for measuring variables

(Adapted from MacRae and Mehuys, 1985.)

Crop-specific attributes need to be considered in residue management. Adapted varieties must be selected to provide sufficient residue in a given climatic zone for erosion control. Consideration needs to be given to dwarf vs. tall varieties based on the amount of residue required. The relationship between harvest index and specific leaf weight is critical to determining the impact of residue on soil erosion. Due to physical differences of plant parts other plant

Table 4. Effect of prior crop on erosion expressed as a percent of moldboard plow

Soil	Prior crop	Chisel plow	No-till
Monona	corn	25	11
	soybean	86	38
Clarion	corn	48	17
	soybean	57	25

(Adapted from Laflen and Colvin, 1982.)

attributes such as shoot:root ratio and stem:leaf ratio will also be critical in residue management decisions.

The quality of crop residue needs to be considered as much as the quantity of residue for soil erosion control. Oschwald and Siemens (1976) found that there was substantially more erosion following soybeans than following corn. The effect of the prior crop on erosion was quantified by Laflen and Colvin (1982) on two Iowa soils shown in Table 4. When soybean was the prior crop, there was more erosion on chisel plow and no-till systems relative to the moldboard plow than when corn was the prior crop. Similar results from different soils suggest soybean affects soil erodibility differently than corn. One possible cause may be the effect of oil seed crops on the surface tension of water. Young (1992, personal communication) using various dilutions of crop residue, showed soybean and sunflower residues decreased the surface tension of water more than corn stover residues. The decrease in surface tension was apparently related to a higher oil content of the residue. These results suggest oil seed crops may affect aggregate stability by decreasing surface tension of water more than other crops. While these results support the explanation for more erosion following soybean, further work is needed to understand the microbial interactions that result from the difference in residue oil and protein content, carbon:nitrogen (C:N) ratio and the decrease in surface tension.

One aspect of residue management often not considered is the amount of water required by the crop to produce the given amount of residue. Generally, there is a unique linear relationship between water transpired by plants and biomass produced (Hanks, 1983; Power et al., 1961; Bauder et al., 1978; Stewart et al., 1977). The strong linear relationship between biomass yield and evapotranspiration (ET) suggests the amount of residue produced is directly related to water available for plant growth. Fertilization can impact the slope of biomass yield vs. ET relationship with higher fertilizer increasing the slope (Power et al., 1961). The water consumed through ET can vary considerably due to location for the same crop species. A few examples of consumptive use for alfalfa, corn, cotton, wheat and soybean are summarized in Table 5 for selected locations in the U.S. The large variation within the same species indicates that ET strongly depends on the evaporative demand at each location.

Table 5. Regional and seasonal differences in water required for crop production.

Crop	Location	ET (mm)	Reference
Alfalfa	Oakes, ND	642	Benz et al., 1984
	Kimberly, ID	1016	Wright, 1988
Corn	Carrington, ND	463	Stegman, 1982
	Bushland, TX	721	Musick and Dusek, 1980
Cotton	Firebaugh, CA	607	Wallender et al., 1979
	Maricopa, AZ	915	Bucks et al., 1988
Wheat	Logan, UT	410	Rasmussen and Hanks, 1978
	Kimberly, ID	648	Jensen et al., 1990
Soybean	Manhattan, KS	320	Clawson et al., 1976
	Manhattan, KS	631	Kanemasu et al., 1976

For soybean, there is approximately a two-fold difference in ET depending on the climatic interactions and evaporative demand for two different years at the same location. In the Great Plains, limited rainfall is often the limiting factor for residue production (Unger et al., 1988; Steiner, 1993).

Doorenbos and Kassam (1979) present a wide-range of water requirements by various crops around the world. Most important in determining this water requirement is the evaporative demand for a specific location. The amount of water required per year for adequate economic growth can range from as little as 400 mm to as much 1600 mm per year for alfalfa. This wide range in water requirement associated with the wide range in economic value for various crops can impact decisions on residue management and maximum economic yield.

Where residue is important for conserving soil water and minimizing wind and water erosion, the efficiency of residue production per unit of water transpired can be the determining factor. An example of the water required for biomass production in Kansas is summarized in Table 6 from Hattendorf et al. (1988) that shows grain and biomass yield of selected crops along with average water use. While the crops are generally planted for grain yield, it is interesting to note residue produced per unit of water transpired. Millet [*Pennisetum americanum* (L.) Leeke], corn and sorghum [*Sorghum bicolor* (L.) Moench] produce relatively larger amounts of residue per unit of water compared to soybean [*Glycine Max* (L.) Merr] and pinto bean (*Phaseolus vulgaris* L.). Thus, where water is limiting and maximum grain production is desired, species should be selected for adequate residue production per unit of water used.

Historically, residue management has been concerned with above-ground residue with little attention given the plant root system as subsurface residue. The plant root system is difficult to quantify but provides for water and nutrient extraction, a source of growth regulators for the top and anchorage for surface biomass. The plant root system can be used to sequester carbon and redistribute

Table 6. Water required for grain and residue production in Kansas

Crop	Grain yield	Total biomass yield	Season water use	Residue/water used
	(kg ha^{-1})	(Mg ha^{-1})	(mm)	(Mg ha^{-1} m^{-1})
Corn	7551	20.1	565	22.2
Sorghum	6277	15.7	484	19.5
Millet	2415	15.6	489	26.8
Pinto bean	2077	6.8	424	11.1
Soybean	2982	8.7	541	10.6
Sunflower	2275	11.2	545	16.3

Average for two years and two locations

(Adapted from Hattendorf et al., 1988.)

it through the soil profile. Organic C, deep within the soil profile as a result of deep rooting, tends to be protected and less susceptible to decomposition than C near the surface. Root exudates are important for maintaining aggregate stability.

One aspect of roots not well understood is the effect of root exudates and residue on soil properties. The potential impact of roots was illustrated by Crookston and Kurle (1989) who evaluated removal of above-ground residue on rotational effects of corn and soybean. They found a significant effect of the previous crop (rotation effect) on both corn and soybean yield but that the removal or addition of surface residue had no effect on the yield of either crop. The yield response of corn and soybean to the rotation was apparently due to the root systems. The economic benefit of rotations was substantiated by Crookston et al. (1991) who showed annually rotated corn and soybean yielded 10% and 8% better, respectively, than monoculture crops.

Genetic variability in the rooting depth for several species has been reviewed by O'Toole and Bland (1987). Kaspar et al. (1978) showed maximum soybean rooting depth at maturity ranged from 1.7-2.0 m. The additional rooting depth can mean additional water to allow the plants to subsist during drought stress. More important is the variation in water use efficiency and shoot:root ratio illustrated for cotton (Quisenberry and McMichael, 1991; McMichael and Quisenberry, 1991) summarized in Table 7. The interspecific variation shows slight differences in water use efficiency with larger differences in the root:shoot ratio. A three-fold change in the shoot:root ratio suggests genetic potential to modify root biomass. These results need testing under field conditions to demonstrate that economic yield can be maintained with the increase in the shoot:root ratio for total residue management.

One of the key elements in crop residue management is carbon cycling. Photosynthesis fixes the carbon and incorporates it into biomass and grain. Tillage and soil management partially or completely incorporate the residue so microbial decomposition and respiration recycles the carbon dioxide back to the

Table 7. Interspecific variation in roots, shoots, and water use efficiency of selected cotton genotypes

Genotype	Shoot biomass	Root biomass	Total water use	Water use efficiency	Root:shoot ratio
	(g plant^{-1})	(g plant^{-1})	(kg plant^{-1})	(g kg^{-1})	
T184	13.03	3.95	8.84	1.93	0.304
T115	13.61	2.79	9.10	1.82	0.209
T80	17.43	2.45	9.00	2.22	0.143
T45	16.41	2.36	9.16	2.07	0.151
T169	14.78	1.87	8.86	1.88	0.126
L. Dwarf	14.27	1.63	9.11	1.76	0.116

(From Quisenberry and McMichael, 1991; McMichael and Quisenberry, 1991.)

atmosphere. Recycling of carbon involves several different cycles within agricultural ecosystems as illustrated in Figure 1. The interaction of the carbon, water and nitrogen cycles plays a large role in the stability of soil organic matter and residue maintenance for erosion control. The complex interactions vary dynamically in time and space and depend on components within these three cycles. The C:N ratio of plant material plays an intimate role. Water required for total biomass production and as the mode of transport for chemicals within the three cycles is critical to understanding and enhancement of residue management. The transformations and transport processes within each of these cycles are only partially understood. The importance of water as it affects biological activity and as a transport agent makes soil water the key factor in carbon and nutrient cycles.

IV. Climate

Climate plays a primary role in residue amounts and quality through two main factors, water and temperature. Rainfall distribution within the U.S. can range from as low as 100 mm per year in the deserts to over 2540 mm per year in the mountains along the West Coast. In agricultural production areas, average annual rainfall can range from 300 mm on the west edge of the Corn Belt to as much as 1170 mm on the east edge. The spatial distribution and timing of the rainfall show wide variation in amounts and distribution during the growing season which requires critical management for maximum production efficiency during climate extremes.

The spatial variability of the climatic factors in the U.S. is demonstrated in Figure 2 using the average length of the frost free period (USDA, 1941). The frost free period ranges from over 320 days in southern Florida to less than 100 days in northern Minnesota with irregular patterns. The irregular boundaries for

CYCLES IN AGRICULTURAL ECOSYSTEMS

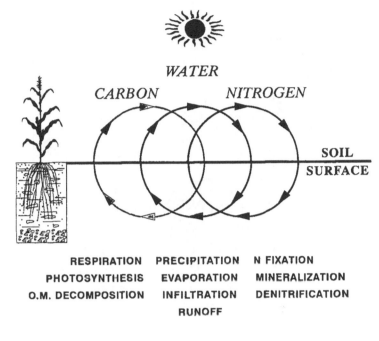

RESPIRATION	PRECIPITATION	N FIXATION
PHOTOSYNTHESIS	EVAPORATION	MINERALIZATION
O.M. DECOMPOSITION	INFILTRATION	DENITRIFICATION
	RUNOFF	

Figure 1. Interactive relationships of the carbon, water and nitrogen cycles in agricultural ecosystems.

the frost-free periods do not parallel the political boundaries and present a challenge for residue management over large areas. Combining spatial variability of the frost-free period with similar spatial variability for rainfall amounts and distribution, temperature extremes and potential evaporation yields a very complex system that requires site-specific data for proper management of crop residue.

Even more critical is the year-to-year variation in climatic extremes at a given location. The long-term air temperature data for Morris, Minnesota (S. D. Evans, West Central Experiment Station, 1992, personal communication) in Figure 3 illustrates the 106 year daily average maximum and minimum temperature and the record extremes for each day. The average maximum and minimum shows typical sinusoidal annual curve related to solar radiation. The maximum and minimum recorded values are as much as 23°C above or below the average maximum or minimum during the winter and as much as 15°C above or below the average maximum or minimum during the summer. The wide range in temperature extremes and the controlling influence of temperature on all of the biological processes related to residue production present a real management challenge. While the historical long-term averages provide guidelines, more

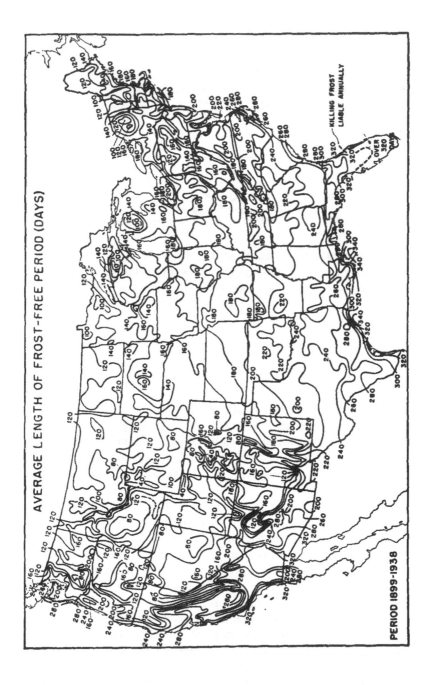

Figure 2. Spatial variability of the frost-free period over the U.S. (From USDA, 1941.)

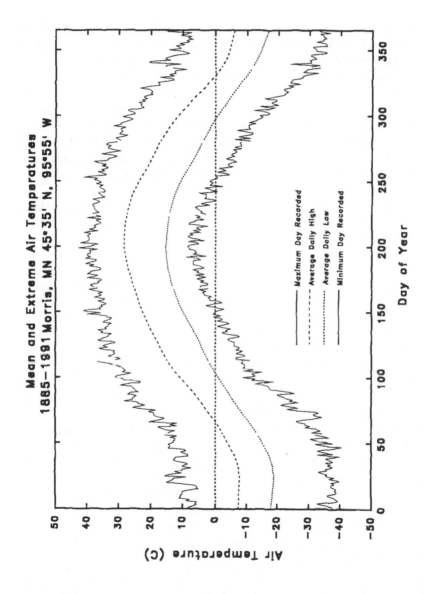

Figure 3. Long-term daily mean and record maximum and minimum air temperatures at Morris, Minnesota.

information is needed on how to cope with the random extreme values that deviate from the mean. A practical compromise between long-term averages vs. point measurements in space and time is needed.

V. Management

The complex interactions of the soil, the crop and the climate as they affect both grain and non-grain yields for residue and erosion control, will require higher levels of management. There are many regional differences in how producers attempt to manage crop residue for soil erosion control and water conservation. Regionalization is one way to organize the information, however it does provide some constraints on how the information is accepted and utilized by the farmers. Site-specific technical information for local problem-solving that carries through to large area basins controlled by political decisions is essential. There is a critical need for efficient technology transfer for improved residue management decisions from site specific soils through the various physiographic or ecological and political levels.

Conservation production systems must utilize residue management, reduce tillage, and minimize evaporation losses. These management practices change along environmental gradients within each of the crop production areas. The major driving variables are economics, tillage, crop management and the changing climatic patterns within each production area. A conceptual framework for handling the tremendous amounts of data and analysis required in evaluating agricultural ecosystems was suggested by Elliot and Cole (1989) and Stewart and Elliot (1991) for the Great Plains and accounts for the long-term ecological changes not easy to predict or measure. Simulation models and large scale data bases organized with a Geographical Information System (GIS), enable evaluation of the current status of the agro-ecosystems and the effect of new long-term management practices on production. While much of their analysis was on the Great Plains, a similar analysis specific for residue management is needed for other physiographic regions.

One example of coping with the complex interactions of the soil, climate and crop is geographic subdivision presented by Allmaras et al. (1991). They define 11 Tillage Management Regions (TMR) for aggregation of survey data at the county level to reflect soil, climate, adopted crop and cultural factors with respect to specific conservation tillage systems. Each TMR has typical soils and climate that restrict crops to selected species that would benefit from GIS and crop growth models. The soil crop tillage interactions were unique to each area within the climate constraints. This organizational structure is a step in the right direction for handling information on large areas.

Inclusion of conservation tillage within a TMR is important for development of improved tillage and residue management systems. The challenge is to identify meaningful indicators of residue management and couple these with efficient spatial assessment techniques such as GIS. Spatial information systems are designed to contain, portray and analyze information within the spatial context. Several examples of geographically referenced information delivery systems are discussed by Anderson and Hanson (1991). They are developing a spatial modelling tool called GRIDS (Geographically Referenced Information Delivery System) which is capable of accepting and outputting georeference data within a spatial framework. The utilization of spatial technology and simulation modelling can be used for providing data bases and management decision aids. GRIDS will have the capability of integrating modelling output and remote sensing data from a geographic information system called GRASS (Geographical Resource Analysis Support System) and an image processing package called ERDAS (Earth Resources Data Analysis System). Combining the spatial technology with the system modelling will allow simulation of system processes within a spatial and temporal framework useful in making management decisions. With further resources and research for residue management, this approach will be beneficial for efficient erosion control over large areas.

One important requirement is availability of data bases that allow site-specific technical decisions for given soil type with data transferability through various political units up to the largest geographic area that could include Major Land Resource Areas (MLRA) as defined by the SCS (USDA-SCS, 1981) or Ecoregions as defined by EPA (Omernik, 1987; Omernik and Gallant, 1990; Gallant et al., 1989). Selection of the major organizational unit may seem arbitrary with MLRA's or Ecoregions being equally suitable. Preliminary observations suggest some advantage for the ecoregion concept because regional variations and causal factors can be integrated within such a system. Major Land Resource Areas consist of geographically associated land resource units (USDA-SCS, 1981) used to organize available information on land resources for farming, ranching, forestry engineering, recreation and other uses. The dominant physical characteristics of each land resource region are described under five headings: (1) Land Use, (2) Elevation and Topography, (3) Climate, (4) Water, and (5) Soil. The data include ranges of parameters and long-term averages that may be useful in residue management. The emphasis on soils data already available enables the MLRA concept to serve as the foundation of a geographically based system upon which enhancements for improved residue management can be added as resources and new information becomes available.

The ecosystem concept integrates soil characteristics, physiography, climatic and plant factors (Gallant et al., 1989). Omernik (1987) states the primary function of an ecoregion map is to provide a geographic framework for organizing the ecosystem resource information. This framework should allow managers, planners and scientists to: (1) compare similarities and differences within the land-water relationships, (2) establish water quality standards that are in tune with regional patterns of tolerances and resiliences to human impacts, (3) locate monitoring and demonstration or reference sites, (4) extrapolate from existing site specific studies, and (5) predict the effects of changes in land use and pollution controls (Omernik, 1987). Resource management interests and priorities of various ecoregions are probably different based on climatic, economic and social constraints.

The key to the successful use of any GIS is the ability to take site specific information and show its up-scale utility as we move to larger geographic areas. A major requirement is a smooth transition from site-specific technical information and detailed modelling on a single farm through various political units to the national level. One possible schematic representation is shown in Figure 4. Each regional classification unit or Residue Management Resource Area (RMRA) would have a set of rules governed by technical factors such as soils, climate, crops, tillage, fertilizer use as well as economic and environmental factors. Major Land Resources Areas, (USDA-SCS, 1981) within each state are politically subdivided through counties and townships to the individual farm level. At this level, a single farm manager would be making the major resource management decisions. The individual farm would likely contain several different field crops and soil types that may require different management. Emphasis is placed on the importance of the transition, continuity and integrity of the data from the individual farm manager, with several different soils, through several political boundaries to the national level where policy decisions are implemented.

One such mechanism specific for residue management illustrated schematically in Figure 5 incorporates several concepts of Stewart and Elliot (1991) for the Great Plains. Combining the spatial technology with the appropriate data bases and process models applied to the scale of interest can result in intelligent management. Critical to intelligent residue management are data bases in Figure 5 that interact with the process models to enable integration of knowledge of soil, physical and biological processes in crop production. Modelling, with the appropriate data bases, can enable management decisions to reduce erosion through improved residue management and can provide a framework for investigating environmental impacts of different management strategies. Simulation models can also be used to examine consequences of different decisions; e.g. utilizing different species under limited water, temperature and nutrients. Combinations of data bases and process models applied to the scale of interest and using real time satellite data will allow up-to-the-minute decisions for cost efficient production while preserving the soil resource. Superimposed over this diagram is the need to balance economic considerations and environmental constraints for long-term sustainability.

One aspect of the comprehensive scheme shown in Figure 5 that needs emphasis is management decisions for the non-crop period. Crop production areas at northern latitudes have less time due to temperature constraints for biological control of environmental impact. It is imperative to manage the crop during the growing season for most efficient and economical benefits by minimizing wind and water erosion and groundwater contamination during the non-growing period. Cover crops that act as "live residue" and scavengers for excess nitrates can be utilized in many production systems to protect the soil during the non-cropping period (Power and Biederbeck, 1991; Vrana, 1991).

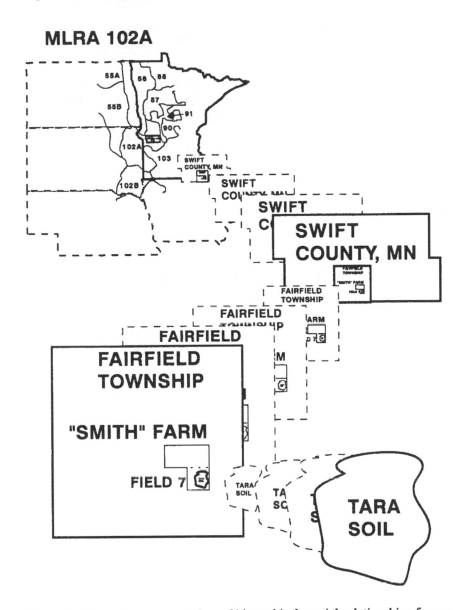

Figure 4. Schematic representation of hierarchical spatial relationships from a specific soil to Major Land Resource Areas.

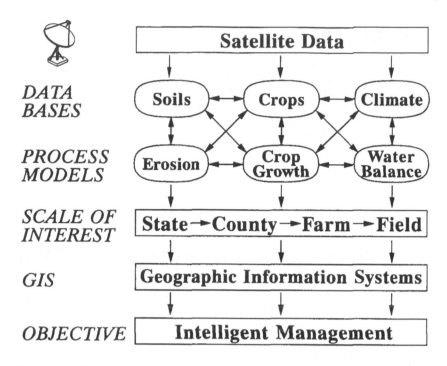

Figure 5. Schematic representation of data bases, process models, scale and real-time data used with geographic information systems for intelligent residue management.

The crop production system must be managed for the entire year rather than just the crop growth period if we are to maintain environmental quality.

The need for realistic simulation models for residue management systems cannot be overstated. With over 12,000 soil mapping units within the U.S. and wide variation in soil physical properties, coupled with climatic differences that range from tropic to subarctic, all require computers and realistic simulation models to cope with the multitude of factors. Crop production using over 100 crops and different cropping systems, that may or may not include irrigation, makes it difficult to develop management systems for all regions on a trial and error basis. Our understanding of the unique and complex interactions of soil, water and temperature on the microbial processes and residue decomposition is critical for improved residue management. The complex interactions of the carbon, nitrogen, and the water cycles with the crop production system, require computers to handle the large data bases and process models. Not only do we have to be concerned about the seasonal dynamics but also the daily dynamics as high soil temperatures can enhance residue decomposition before crop canopy closure. Adding to the complexity is the unknown number of independent managers who farm the soil making management decisions based on economic and global market considerations. Overshadowing the technical aspects of residue management is the potential political intervention as more people become concerned about environmental quality and mandatory control. Accurate computer models can be useful tools to make both management and policy decisions.

VI. Research Needs for Improved Residue Management

Research needs unique to the complex interactions of soil, plant and climate, on residue management follow.

Short-term research (i.e. 3 years) is needed to cope with spatial variability within a field. The incorporation of both spatial and temporal variation in the process models and systems research will aid in management decisions. Techniques to enable interpolation/extrapolation of existing data to other field locations are required. The farming by soil concept is one step to maximize production efficiency. Improved plant growth and erosion process models are needed to enhance the quality of technical and political decisions. The incorporation of improved data bases into standardized GIS and other regional systems that aggregate and allow georeferencing of resource data will allow improved management decisions.

New tillage and cropping systems with long-term non-grain management in mind are needed. The interaction of tillage with the carbon, nitrogen and water cycles as they affect residue longevity in different climatic regimes needs further research. Improved equipment for residue management and partial incorporation in selected situations is necessary. New cropping sequences are needed to preserve and maintain the soil resource during the non-growing period.

Long-term research (i.e. 20 years) is needed to select species and varieties for stable non-grain yield (biomass or residue) with minimum allelopathic effects as well as economic grain yield. The non-grain yield should be designed and managed specifically for off-season erosion and nutrient-release control. Species selection with variable shoot:root ratio that incorporates aspects of the dwarf vs. tall-growing varieties is required for maximum alternatives for residue management. Varieties and species need to be selected based on root characteristics that affect aggregate stability as well as sequester carbon to greater root depths.

Cold-tolerant species as economic cover crops or a source of residue are needed in northern climates. New and improved cover crops and cropping sequences are needed that provide ground cover for erosion control and still allow timely recharge of soil water. Research is needed on understanding the basic soil, plant, water relationships as they impact both quantity and quality of residue. The genetic diversity for protein content and C:N ratio as they impact the rate of residue decomposition on the soil surface needs to be identified. The critical soil and crop biochemical interactions of organic matter decomposition and aggregate stability need clarification. The influence of oil seed crops on soil erosion, the role of various chemicals and fertilizers on microorganisms and soil fauna as they impact residue management and mineralization needs attention. Research is needed to determine the impact of actual root depth for a given soil relative to the potential root depth under ideal soil conditions for increasing rooted volume and water extraction. Root function for water and nutrient extraction for biomass production requires further research to increase both economic yield and production efficiency.

Research is needed to enhance soils and climate data bases to develop improved climate generators. Further research is needed to manage site-specific extreme erosive events rather than utilizing long-term broad-area averages. Understanding catastrophic events is needed to develop and implement new technology for intelligent residue management.

Detailed information is needed on hydraulic characteristics for infiltration and unsaturated flow within the plant root zone. Water balance models require unsaturated conductivity as a function of water content as affected by bulk density, compaction and various methods of tillage and residue incorporation before they are sufficiently accurate for site-specific recommendations. The interaction between soil properties, residue properties, tillage practices, water infiltration, evaporation, storage and evapotranspiration over a wide range of climatic conditions and soil types is only partially understood. While some basic principles are understood, additional research is needed to extrapolate site-specific information to develop flexible management practices across U.S. production areas.

The above research will result in more effective and widely-adapted conservation tillage systems with improved residue management.

References

Allmaras, R.R., G.W. Langdale, P.W. Unger, R.H. Dowdy and D.M. VanDoren. 1991. Adoption of conservation tillage and associated planning systems. p. 53-83. In: R. Lal and F.J. Pierce (eds.) *Soil Management for Sustainability*. SWCS, Ankeny, IA.

Anderson, G.L. and J.D. Hanson. 1991. A Geographically Referenced Information Delivery System. p. 99-108. In: J.D. Hanson, M.J. Schaeffer, D.A. Ball, and C.V. Cole (eds.) *Sustainable Agriculture for the Great Plains*. Symposium Proc., USDA-ARS. 255 pp. Sept. 1989.

Bauder, J.W., A. Bauer, J.M. Ramirez, and D.K. Cassel. 1978. Alfalfa water use and production on dry land in the irrigated sandy loam. *Agron. J.* 70(1):95-100.

Benoit, G.R. 1973. Effect of freeze thaw cycles on aggregate stability and hydraulic conductivity of three soil aggregate sizes. *Soil Sci. Soc. Am. Proc.* 37(1):3-5.

Benoit, G.R., S. Mostaghimi, R.A. Young, and M.J. Lindstrom. 1986. Tillage residue effects on snow cover, soil, water, temperature and frost. *Trans. ASAE* 29(2):473-479.

Benz, L.C., E.J. Doering, and G.A.Reichman. 1984. Water-table contribution to alfalfa evapotranspiration and yields in sandy soils. *Trans. ASAE* 27:1307-1312.

Borg, H. and D. W. Grimes. 1986. Depth development of roots with time: an empirical description. *Trans. ASAE* 29(1):194-197.

Brown, D.A. and H.D. Scott. 1984. Dependency of crop growth and yield on root development and activity. p. 101-136. In: S.A. Barber and D.R. Bouldin (eds.) *Roots, Nutrient and Water Influx, and Plant Growth.* Spec. Publ. No. 49, ASA, Madison, WI.

Bucks, D.A., S.G. Allen, R.L. Roth, and B.R. Gardner. 1988. Short staple cotton under micro and level irrigation methods. *Irrig. Sci.* 9:161-176.

Clawson, K.L., J.E. Specht, B.L. Blad, and A.F. Garay. 1986. Water use efficiency in soybean pubescence density isolines - A calculation procedure for estimating daily values. *Agron. J.* 78:483-487.

Crookston, R.K. and J.E. Kurle. 1989. Corn residue effect on the yield of corn and soybean grown in rotation. *Agron. J.* 81(2):229-232.

Crookston, R.K., J.E. Kurle, P.J. Copeland, J.H. Ford, and W.E. Lueschen. 1991. Rotational cropping sequence effects yield of corn and soybean. *Agron. J.* 83(1):108-113.

Doorenbos, J. and A.H. Kassam. 1979. Y*ield response to water.* FAO Irrigation and Drainage Paper, No. 33. FAO, Rome, Italy. 193 pp.

Doorenbos, J. and W.O. Pruitt. 1977. *Guidelines for prediction of crop water requirements.* FAO Irrigation and Drainage Paper, No. 24. FAO, Rome, Italy. 179 pp.

Edwards, W.M. and L.B. Owens. 1991. Large storm effects on total soil erosion. *J. Soil Water Conserv.* 46(1):75-78.

Elliot, E.T. and C.V. Cole. 1989. A perspective on agro-ecosystem science. *Ecology* 70(6):1597-1602.

Fehrenbacher, J.B. and R.H. Rust. 1956. Corn root penetration in soils derived from various textures of Wisconsin-age glacial till. *Soil Sci.* 82:369-378.

Gallant, A.L., T.R. Whittier, D.P. Larsen, J.M. Omernik, and R.M. Hughes. 1989. *Regionalization as a tool for managing environmental resources.* NSI Technology Services Corporation, USEPA 600/3-89/060 Environmental Research Laboratory, Corvallis, Oregon.

Hamblin, A.P. 1985. The influence of soil structure on water movement, crop root growth and water uptake. *Adv. Agron.* 38:95-158.

Hanks, R.J. 1983. Yield and water use relationships: an overview. p. 393-411. In: H. M. Taylor, W.R. Jordan, and T.R. Sinclair (eds.) *Limitations to efficient water use in crop production.* ASA, Madison, WI.

Hattendorf, M.J., M.S. Redelfs, B. Amos, L.R. Stone, and R.E. Gwin, Jr. 1988. Comparative water use characteristics of six row crops. *Agron. J.* 80(1):80-86.

Jensen, M.E., R.D. Burman, and R.G. Allen. 1990. ASCE manuals and reports on engineering practices No. 70, table 6.14, 154. In: M.E. Jensen, R.D. Burman, and R.G. Allen (eds.) *Evapotranspiration and Irrigation Water Requirements*. 332 pp. ASCE, New York, New York.

Kanemasu, E.T., L.R. Stone, and W.L. Powers. 1976. Evapotranspiration Model Tested for Soybean and Sorghum. *Agron. J.* 68:569-572.

Kaspar, T.C., C.D. Stanley, and H.M. Taylor. 1978. Soybean root growth during the reproductive stages of development. *Agron. J.* 70(6):1105-1107.

Laflen, J.M. and T.S. Colvin. 1982. Soil and water loss from no-till narrow row soybeans. Paper No. 82-2023. Presented at the 1982 Summer Meetings of the American Society of Agricultural Engineers. June 27-30, 1982. University of Wisconsin at Madison, WI.

MacRae, R.J. and G.R. Mehuys. 1985. The effect of green manuring on the physical properties of temperate area soils. *Adv. Soil Sci.* 3:71-94.

McCoy, E.L. 1987. Energy requirements for root penetration of compacted soil. p. 367-377. In: L.L. Boersma (ed.) *Future Developments in Soil Science Research*. Soil Sci. Soc. Am., Madison, WI.

McMichael, B.L. and J.E. Quisenberry. 1991. Genetic variation for root-shoot relationships among cotton germplasm. *Environ. Experimental Bot.* 31(4):461-470.

Musick, J.T. and D.A. Dusek. 1980. Irrigated corn yield response to water. *Trans. ASAE* 23:92-98, 103.

O'Toole, J.C. and W.L. Bland. (1987). Genotypic variation in crop plant root systems. *Adv. Agron.* 41:91-145.

Omernik, J.M. 1987. Ecoregions of the conterminous United States. *Ann. Assn. Am. Geog.* 77(1):118-125.

Omernik, J.M. and A.L. Gallant. 1990. Defining regions for evaluating environmental resources. p. 936-947. In: *Global Natural Resource Monitoring and Assessments: Preparing for the 21st century*. Vol. 2. American Society of Photogrametry Remote Sensing. Bethesda, MD.

Oschwald, W.R. and J.C. Siemens. 1976. Soil erosion after soybeans. p. 74-81. In: Lowell D. Hill (ed.) *World Soybean Research*. Interstate Printers and Publishers, Inc., Danville, IL.

Power, J.F. and V.O. Biederbeck. 1991. Role of cover crops in integrated crop production systems. p. 167-174. In: W. L. Hargrove (ed.) *Cover Crops for Clean Water*. Soil and Water Cons. Soc., Ankeny, IA.

Power, J.F., D.L. Grunes, and G.A. Reichmann. 1961. The influence of phosphorus fertilization and moisture on growth and nutrient absorption by spring wheat: I. Plant growth, N uptake and moisture use. *Soil Sci. Soc. Am. Proc.*, 25:207-210.

Quisenberry, J.E. and B.L. McMichael. 1991. Genetic variation among cotton germplasm for water-use efficiency. *Environ. Experimental Bot.* 31(4):453-460.

Rasmussen, V.P. and R.J. Hanks. 1978. Spring wheat yield model for limited moisture conditions. *Agron. J.* 70:940-944.

Smika, D.E. and C.J. Whitfield. 1966. Effect of standing wheat stubble on storage of winter precipitation. *J. Soil Water Conserv.* 21:138-141.

Snyder, J.R., M.D. Skold, and W.O. Willis. 1979. The economics of snow management: An application of game theory. *Western J. Agric. Econ.* 4(2):61-71.

Soil Conservation Service. 1967. *The national inventory of soil and water conservation needs.* USDA Statistical Bull., No. 461.

Stegman, E.C. 1982. Corn yield as influenced by timing of evapotranspiration deficits. *Irrig. Sci.* 3:75-87.

Steiner, J.L. 1993. Crop residue effects on water conservation. In: P.W. Unger (ed.) *Managing agricultural residues.* CRC Press Inc., Boca Raton, FL (in press).

Stewart, J.I., R.E. Danielson, R.J. Hanks, E.B. Jackson, R.M. Hagen, W.O. Pruitt, W.T. Franklin, and J.P. Riley. 1977. *Optimizing crop production through control of water and salinity levels in the soil.* Utah Water Res. Lab PR 151-1, Utah State Univ., Logan. 191 pages.

Stewart, J.W.B. and E.T. Elliot. 1991. A conceptual framework for regional analysis for semi-arid to sub-humid agro-ecosystems. p. 17-30. In: J.D. Hanson, M.J. Schaeffer, D.A. Ball, and C.V. Cole (eds.) *Sustainable Agriculture for the Great Plains.* Symposium Proc., USDA-ARS, ARS-89, 255 pp.

Tennant, D. 1976. Wheat root penetration and total available water in a range of soil types. *Aust. J. Exp. Agric. Anim. Husb.* 16:570-577.

Unger, P.W., G.W. Langdale, and R.I. Papendick. 1988. Role of crop residues - improving water conservation and use. p. 69-100. In: *Cropping Strategies for Efficient Use of Water and Nitrogen.* Spec. Publ. 51. ASA, Madison, WI.

U.S. Department of Agriculture - Soil Conservation Service. 1981. *Land resource regions and major land resource areas of the United States.* USDA-SCS Agric. Handb. 296. U.S. Gov. Print. Office, Washington, D.C.

U.S. Department of Agriculture, Yearbook of Agriculture. 1941. Page 746.

Van Doren, D.M. Jr. and R.R. Allmaras. 1978. Effect of residue management practices on the soil physical environment, microclimate and plant growth. p. 49-83. In: W.R. Oschwald (ed.) *Crop Residue Management Systems.* ASA Spec. Publ. 31.

Vrana, V.K. 1991. Crop residue management for conservation. Proceedings of a National Conference held August 8 & 9, 1991, Lexington, KY. SWCS, Ankeny, IA.

Wallender, W.W., D.W. Grimes, D.W. Henderson, and L.K. Stromberg. 1979. Estimating the contribution of a perched water table to the seasonal evapotranspiration of cotton. *Agron. J.* 71:1056-1060.

Wischmeier, W.H. 1973. Conservation tillage to control water erosion. In: Conservation Tillage. *Proceedings of a National Conference.* p. 133-141. SCSA, Ankeny, IA.

Wright, J.L. 1988. Daily and seasonal evapotranspiration and yield of irrigated alfalfa in southern Idaho. *Agron. J.* 80:662-669.

Young, R.A., M.J. Römkens, and D.K. McCool. 1990. Temporal variations in soil erodibility. p. 41-53. In: Bryan B. Rorke (ed.) *Soil erosion experiments and models*. Catena Supplement No. 17. Cremlingen - Destedt, West Germany.

Index